Traditional Herbal Remedies of Sri Lanka

Natural Products Chemistry of Global Plants

Editor-in-Chief
Raymond Cooper, PhD

This unique book series focuses on the natural products chemistry of botanical medicines from different countries such as Turkey, Sri Lanka, Bangladesh, Vietnam, Cambodia, Brazil, China, Borneo, Thailand, and African and Silk Road countries. These fascinating volumes are written by experts from their respective countries. The series will focus on the pharmacognosy, covering recognized areas rich in folklore as well as botanical medicinal uses as a platform to present the natural products and organic chemistry. Where possible, the authors will link these molecules to pharmacological modes of action. The series intends to trace a route through history from ancient civilizations to the modern day showing the importance to man of natural products in medicines, foods, and a variety of other ways.

Recent Titles in this Series

Traditional Herbal Remedies of Sri Lanka, *Viduranga Y. Waisundara*

Traditional Herbal Remedies of Sri Lanka

Viduranga Y. Waisundara

CRC Press is an imprint of the
Taylor & Francis Group, an **informa** business

CRC Press
Taylor & Francis Group
6000 Broken Sound Parkway NW, Suite 300
Boca Raton, FL 33487-2742

First issued in paperback 2021

ISBN-13: 978-0-367-78009-8 (pbk)
ISBN-13: 978-1-138-74308-3 (hbk)

Library of Congress Cataloging-in-Publication Data

Names: Waisundara, Viduranga Y., author.
Title: Traditional herbal remedies of Sri Lanka / Viduranga Y. Waisundara.
Description: Boca Raton, Florida : CRC Press, 2019. | Series: Natural products chemistry of global plants | Includes bibliographical references and index.
Identifiers: LCCN 2018057996| ISBN 9781138743083 (hardback : alk. paper) | ISBN 9781315181844 (ebook)
Subjects: LCSH: Medicinal plants--Sri Lanka. | Materia medica, Vegetable--Sri Lanka. | Herbs--Therapeutic use--Sri Lanka. | Traditional medicine--Sri Lanka.
Classification: LCC QK99.S78 W35 2019 | DDC 581.6/34095493--dc23
LC record available at https://lccn.loc.gov/2018057996

Visit the Taylor & Francis Web site at
http://www.taylorandfrancis.com

and the CRC Press Web site at
http://www.crcpress.com

Dedication

To Prof. Ananda Wijekoon and Mr. Lakshman Wijekoon who opened my eyes to the holistic nature of the traditional medicinal system of Sri Lanka

To Ms. Rucheera Jayatunga who inspired me to look for peace within and live in the present moment

Viduranga Y. Waisundara

Contents

Introduction to the Book Series

NATURAL PRODUCTS CHEMISTRY OF GLOBAL PLANTS

CRC Press is publishing a new book series on the *Natural Products Chemistry of Global Plants*. This new series will focus on pharmacognosy, covering recognized areas rich in folklore with botanical and medicinal uses as a platform to present the natural products and organic chemistry and, where possible, link these molecules to pharmacological modes of action. This book series covers the botanical medicines from different countries, including but not limited to Bangladesh, Borneo, Brazil, Cambodia, Cameroon, Silk Road Countries, South Africa, Sri Lanka, Thailand, Turkey, Uganda, Vietnam, and Yunnan Province (China), written by experts from each country. The intention is to provide a platform to bring forward information from under-represented regions.

Medicinal plants are an important part of human history, culture, and tradition. Plants have been used for medicinal purposes for thousands of years. Anecdotal and traditional wisdom concerning the use of botanical compounds is documented in the rich histories of traditional medicines. Many medicinal plants, spices, and perfumes changed the world through their impact on civilization, trade, and conquest. Folk medicine is commonly characterized by the application of simple indigenous remedies. People who use traditional remedies may not understand in Western terms the scientific rationale for why they work but know from personal experience that some plants can be highly effective.

This series provides rich sources of information from each region. An intention of the series of books is to trace a route through history from ancient civilizations to the modern day, showing the important value to humankind of natural products in medicines, foods, and many other ways. Many of the extracts are today associated with important drugs, nutritional products, beverages, perfumes, cosmetics, and pigments, which will be highlighted.

The books will be written for both chemistry students who are at university level and for scholars wishing to broaden their knowledge in pharmacognosy. Through examples of the chosen botanicals, herbs, and plants, the series will describe the key natural products and their extracts with emphasis upon sources, an appreciation of these complex molecules, and applications in science.

In this series, the chemistry and structure of many substances from each region will be presented and explored. Often, books describing folklore medicine do not describe the rich chemistry or the complexity of the natural products and their respective biosynthetic building blocks. By drawing on the chemistry of these functional groups to show how they influence the chemical behavior of the building blocks, which make up large and complex natural products, the story becomes more fascinating. Where possible, it will be advantageous to describe the pharmacological nature of these natural products.

Raymond Cooper
Department of Applied Biology and Chemical Technology
The Hong Kong Polytechnic University
Hong Kong

Prologue

Sri Lanka has a well-established Ayurvedic medicinal system that appears to span several millennia. Records show traditional medicinal practices dating all the way back to the times of King Ravana, who ruled over the country nearly 3,000 years ago. Since ancient times, the Sri Lankan medicinal system has predominantly utilized herbs and spices or their combinations for the treatment of various ailments. This is mostly due to Sri Lanka's status as a tropical country and a biodiversity 'hotspot', gifted with a plethora of flora and fauna. The proposed book will look at the traditional medicinal practices of Sri Lanka, which utilize plant material from a cultural, philosophical, and most importantly, scientific perspective. While the former two aspects were the dominant factors that led to the popularity and comparatively higher utilization of the Sri Lankan traditional medicinal system in ancient times, from a more modern perspective, the scientific facts have influenced the decision-making process when selecting an appropriate medical practitioner for ailments and diseases. In Sri Lanka, the medical practitioners continue to follow the ancient *Ola Leaf*-based texts for administration of medicines. *Ola Leaf* is basically the most primitive form of manuscript that was used particularly in southern India and Sri Lanka to write texts. Due to its location and influence on Ayurvedic medicine, Sri Lanka has benefitted from this confluence of its geography and cultural influences.

When it comes to the scientific aspects, several Sri Lankan herbs have been in the spotlight for possessing novel bioactive constituents that have promising therapeutic effects. Although most of these findings have not progressed beyond this stage of discovery and existence into the phase of formulation of pharmaceuticals, it is hoped that some of the herbs and bioactive constituents that have been under scientific scrutiny will eventually be considered as strong candidates to combat currently prevailing global disease conditions.

This book is structured into three defined sections. First, there is an introduction to Sri Lanka, considered the 'Pearl of the Indian Ocean' due to its location at the southern tip of India, with its traditional medicinal system of Sri Lanka and the traditional medicinal practices in modern times. Second, the causes and prescribed herbal remedies of some common disease conditions are described with many examples. Third, where available, scientific evidence on the therapeutic properties of some popular herbs is described.

Importantly, this book presents the folklore, the selection of plants, and the importance of preparation and weaves all these aspects into the modern findings for each herb and plant described, with recommendations for further scientific and clinical studies where appropriate.

Author

Dr. Viduranga Waisundara earned a PhD in food science and technology from the Department of Chemistry, National University of Singapore, in 2010. She was a lecturer at Temasek Polytechnic, Singapore, from July 2009 to March 2013 after which she relocated to her motherland of Sri Lanka and spearheaded the Functional Food Product Development Project at the National Institute of Fundamental Studies until October 2016. Dr. Waisundara subsequently became a senior lecturer on a temporary basis in the Department of Food Technology, Faculty of Technology, Rajarata University of Sri Lanka until July 2018. She is presently Coordinator of the science programs and Deputy Principal of the Australian College of Business and Technology – Kandy Campus, in Sri Lanka.

Dr. Waisundara currently serves as the Global Harmonization Initiative (GHI) Ambassador to Sri Lanka and has been the co-chair of the GHI Working Group on Nutrition since November 2018. She has played an active role in spreading the word about harmonizing food safety regulations based on scientific facts and improving food security. She has been an invited speaker in international conferences and participated in local school events in Sri Lanka to spread awareness on functional food and dietary habits. Together with these commitments, Dr. Waisundara has also contributed to one of the recent initiatives of the National Education Commission of Sri Lanka to evaluate postgraduate research in the country for formulating policies and regulations.

Dr. Waisundara has edited/co-edited the following books by InTech Publishers, Croatia: *Superfood and Functional Food: Development of Superfood and their Roles in Medicine*; *Super Food and Functional Food: An Overview of their Processing and Utilization*; *Cassava*; *Diabetes Food Plan*; *Palm Oil*; and *Biochemistry of Fatty Acids*. She also writes poetry, and her poems on scientific matters are available in the GHI Newsletter (August 2016) and on the Society for Redox Biology and Medicine (SfRBM), USA – Women in Science webpage.

Dr. Waisundara serves on the editorial boards of the following journals: Asian Journal of Medical Principles and Clinical Practice; Annual Research & Review in Biology; Frontiers in Nutrition; and Frontiers for Young Minds.

Dr. Waisundara's interest in traditional medicinal systems stemmed from her postgraduate research project in which she worked on assessing the anti-diabetic properties of *Scutellaria baicalensis*, an herb used for the treatment of diabetes in traditional Chinese medicine. The all-inclusive nature of traditional medicines existing around the world are a point of attraction for her, and a concept that she found to occur in the traditional medicinal system of Sri Lanka, as well. Her family has been very supportive of her interest, and she continues to write about traditional medicinal herbs and their scientific basis of application, so as to popularize these medicines globally.

1 Introduction and the Sri Lankan Traditional Medicinal System

SRI LANKA AND ITS TRADITIONAL MEDICINAL SYSTEM

Sri Lanka (previously called Ceylon during colonial times) is an island nation located in the tropics, lying off the southern tip of the Indian subcontinent (see Table 1.1 for a fact file of Sri Lanka). The geographic location of Sri Lanka is shown in Figure 1.1. The country is home to many cultures, languages, and ethnicities. The majority of the country's population belongs to the Sinhalese ethnicity with a large minority of Tamils. Moors, Burghers, Malays, Chinese, and the aboriginal Vedda groups also being part of the country's population. It has an ancient cultural heritage spanning 3,000 years, with evidence of pre-historic human settlements dating back over 125,000 years. Sri Lanka has a strategic location in the southwest of the Bay of Bengal and to the southeast of the Arabian Sea. Due to the deep-water harbors such as Trincomalee (Figure 1.2), it became a key maritime location from the time of the ancient Silk Road to the Maritime Silk Road. As such, the country was known by many traders for its plethora of spices, which were sold to those travelling along these routes and harbors (Siriweera 1994). There was great commercial importance placed upon spices as a trade commodity and Sri Lanka has been identified since ancient times as a hub of high-quality spices that carry medicinal value. The key spices traded during the days of the ancient Silk Road were cinnamon, cardamom, and cloves. Not only were these spices used as flavoring agents, both locally and overseas, they also carried therapeutic properties – an aspect that was known to the traditional medicinal practitioners of Sri Lanka since ancient times. However, it should be noted that the establishment and popularity of the traditional medicinal system of Sri Lanka since ancient times was independent from the commercial exchanges made during the ancient Silk Road days, although it is possible that the recognition of certain spices with medicinal value as an agricultural crop was imparted due to the influence of international traders visiting the country.

Sri Lanka is like many other tropical countries, possessing a wide range of plant species as well as a significant variability of climatic zones. Although the country is relatively small in size, it has the highest biodiversity density in Asia (Mittermeier, Myers and Mittermeier 2000). It consists mostly of flat lands with mountains existing primarily in the south-central part of the country. There is a significant diversity of flora and fauna owing to these geographical and climatic landscapes. There are many trees, which can be converted into wood for domestic use; commonly found in the dry-land forests are satinwood, ebony, ironwood, mahogany, and teak. Conversely, the wet zone consists

TABLE 1.1
Sri Lanka Fact File

Capital	Administrative: Sri Jayawardenepura Kotte Commercial: Colombo
Area	65,610 km^2
Population	22,409,381 (2017 estimate) Annual population growth rate: 1.14% (2011) Birth rate of 17.6 births per 1,000 people (2011) Death rate of 6.2 deaths per 1,000 people (2011)
Climate	Mean temperatures range from 17°C in the hill areas to a maximum of 33°C in other areas close to sea level Average yearly temperatures range from 28°C to 31°C
Flora and Fauna	Lies within the Indomalaya ecozone One of 25 biodiversity hotspots in the world 27% of the 3,210 flowering plants and 22% of the mammals are endemic

Source: Mittermeier, Myers and Mittermeier 2000.

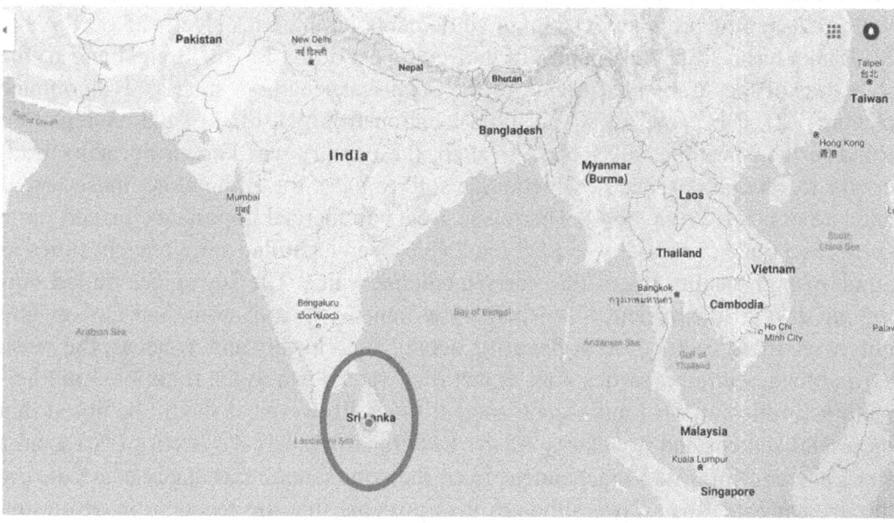

FIGURE 1.1 Geographic location of Sri Lanka on the world map. It is an island located along the southern tip of the Indian sub-continent.

of tropical evergreen forests with tall trees, broad foliage, and a dense undergrowth of vines and creepers. Subtropical evergreen forests resembling those of temperate climates are present in the higher altitudes. Due to the small area of the country, there is easy access to the plants that have medicinal value, since travelling between locations to collect herbs is not all that difficult. Despite continuing deforestation and disappearance of habitat in both the lowlands and hill country, the country still boasts luscious greenery and dense forests that provide accommodation for many unique species of

FIGURE 1.2 Trincomalee – a natural harbor and bay area of maritime interest for more than 2,500 years. It was called 'Gokanna' during ancient times. This point in the country was part of the ancient Silk Road, and in old days, many of the medicinal spices of the country such as cinnamon, cloves, and cardamom were sold to overseas traders in this harbor.

wildlife as well as plants (Figure 1.3). Sri Lanka has two botanical gardens located in Haggala and Peradeniya (Figure 1.4) – both in the central province, where many varieties of plants that are available locally (as well as plants from overseas that were donated as gifts from diplomats) are showcased to the general population.

For generations, the medicinal plants of Sri Lanka have been consumed for the purposes of disease prevention, cures for ailments, or maintaining health and wellness (Waisundara and Watawana 2014a). Owing to the diversity in the flora and fauna, the range of diseases being treated with herbal medicines has been numerous as well. Due to the familiarity of widely available medicinal plants, many Sri Lankans can identify the varieties and types of herbs growing within their area of residence. In fact, the local people have the habit of growing at least one or two medicinal plants in their backyards, or any available space within their area of residence. This acquaintance with the traditional medicines allows the local people to self-medicate. Most of the time, with simple or complex diseases, herbal remedies are consumed without the advice of a traditional medicinal practitioner (Waisundara and Watawana 2014b). The herbs can be consumed in fresh or dried form, depending on the traditional practice or recommendation of the practitioner. Sometimes, the dried forms of herbs are retained and preserved, due to seasonal availability. Nevertheless, the primary reason in the Sri Lankan traditional medicinal system behind making dried forms of herbs was for easing the method of preparation. Most of the herbs used in the Sri Lankan traditional medicinal system are evergreens and are even considered

FIGURE 1.3 Indicators of the vast assortment of flora and fauna available in the country and the collection of remedies they offer for various disease conditions. Evidence of the lush greenery in Sri Lanka: (A) View from Namal Uyana in the north central province – a lowland area, (B) Bible rock and the surrounding foliage as seen from Kadugannawa (hill country), (C) Kuda Dunhinda waterfall surrounded by forest areas located in Badulla – bordering the hill country and lowlands, (D) The misty green plains of Ambewela in the Nuwara Eliya district (hill country). Each view displays a variety of trees and vegetation in terms of their height, color, and growth.

nowadays as weeds. Weeds are generally able to withstand harsh weather conditions and can be found in abundance in any place. In the traditional medicinal system of Sri Lanka, the ability of plants to behave as weeds is in fact seen as a mark of robustness and a characteristic that can impart medicinal effects.

According to local myths and legends, the traditional medicinal system of Sri Lanka goes back 3,000 years or more (Weragoda 1980). It is also called 'Ayurveda', in a similar manner to the term used to refer to the traditional medicinal system of India. At present, and for ease of reference, the traditional medicinal system of Sri Lanka is divided into four main domains: Siddha and Ayurveda, both of which originated in India; Unani, a Perso-Arabic traditional medicine system; and an indigenous medicinal system in Sri Lanka, known as 'Deshiya Chikitsa' (De Zoysa et al. 2017; Ediriweera and Ratnasooriya 2009). The system stresses a balance of three elemental substances associated with heat for achieving physical equilibrium: *āyu* or *vāta* (air/space/wind), *pitta* (fire), and *kapha* (water/earth) (Waisundara, Watawana and Jayawardena 2014). These elements are defined in the Indian Ayurvedic system as well. According to this three-fold dimension, every individual has an innate combination of these elements,

FIGURE 1.4 The Peradeniya Botanical Gardens, located in Kandy, Sri Lanka, where many varieties of local as well as foreign plant species, including an herbal garden, are on display.

for which the required balance requires structuring our behavior or environment (Karalliadde and Gawarammana 2008). Due to the influence of Indian Ayurveda, the elements of *vāta*, *pitta*, and *kapha* remain unchanged when defining diseases in the Sri Lankan traditional medicinal system. However, the two schemes go separate ways when it comes to the treatment methods. The herbs used for the same diseases tend to be different, mostly owing to the matter of availability (Waisundara and Watawana 2014b).

Most of the methods of treatment and remedies have not been systematically documented in the traditional medicinal system of Sri Lanka. The preparation methods of the herbal concoctions and decoctions have been transmitted verbally down through the generations and are administered by traditional medicinal practitioners even to the present day (Ediriweera and Ratnasooriya 2009). Many of these traditional medicinal recipes of Sri Lanka were passed down from one traditional medicinal practitioner to another, usually from father to son, from one generation to another, and if any recipes were written down, they were most commonly preserved in ancient Sinhala. Overall, however, it should be noted that written documents, either original or translations, are scarce. In many instances, there are no fixed formulae for the medicines, since the dosages and combinations were customized according to the physiology of the patients. This is mostly owing to the underlying philosophy of the traditional medicinal system, since it targets the whole body and not each organ in isolation. Typically, the herbal formulations are based on parts obtained from trees and plants, such as the bark, flowers, leaves, fruits, seeds, stems, root, rhizomes, and bulbs (De Zoysa et al. 2017).

THE SRI LANKAN TRADITIONAL MEDICINAL SYSTEM IN MODERN TIMES

The Sri Lankan traditional medicinal system has gradually developed towards approaching the scientific realms for recognition (i.e. evidence based), although the appreciation and consumerism still lies mostly among the local populations (De Zoysa et al. 2017; Ediriweera and Ratnasooriya 2009). In Sri Lanka, about 60%–70% of the rural population still resort to using the traditional Ayurvedic medicines for curing and prevention of diseases (De Zoysa et al. 2017). Most of these medicines are consumed in the form of concoctions or decoctions and methods of use follow preparations developed during ancient times. The local population in Sri Lanka – despite the country being considered as a developing nation – has displayed a significant rate of urbanization together with an increased literacy level and a wider exposure to science and technology. However, it is surprising that given these trends and changes of lifestyle, the traditional Ayurvedic medicines of the country still have a wider acceptance over their more 'Western' counterparts, even though there is a lack of substantial modern scientific evidence to support the claims.

Many of the traditional medicinal practitioners of Sri Lanka have their practices in rural areas, although some of them do travel to urban areas to hold clinics a few days of the week. The practitioners still follow traditional ways: No specific fees are requested from the patients and instead, donations or gifts are accepted. They still follow the ancient *Ola Leaf*-based texts for administration of medicines. *Ola Leaf* is basically the most primitive form of manuscript that was used particularly in southern India and Sri Lanka to write texts. This is simply a dried palm leaf obtained from the talipot tree (*Corypha umbraculifera*). *Ola Leaf*-based writing dates back over 3,000 years, when many sages and fortune tellers wrote peoples' fortunes on the leaves. Most likely originating from the exercise of fortune telling, ancient texts of Ayurveda were written on Ola leaves as well and these texts in turn have been preserved by many of the Sri Lankan Ayurvedic practitioners to date as reference material. Many of the practitioners, however, have another profession, since having an Ayurvedic practice alone may not be profitable. Although most of the practitioners do prepare their own tonics, syrups, pastes, and ointments, for the urban population there is the advantage of going into medicinal halls such as the Kandy Ayurvedic Hall or the Bowatte Traditional Medicinal Hall, which appear to have outlets across the island. As shown in Figure 1.5, many of the herbal medicines are prepared to meet the consumer demands of modern times, in bottles or packets, and they are labelled in English as well. This is so that most of the consumers might be familiar with the English term used for the herb, anticipating tourist trade. Obtaining medicines from these halls is as simple as going to a supermarket for grocery shopping, as these are medicines for uncomplicated diseases such as coughs, colds, and fever. A prescription is not a necessity most of the time to obtain medicines from these medicinal halls, although for complicated herbal formulations the patient wisely obtains a written note, a 'prescription' from the traditional medicinal doctor. For complex diseases, though, especially those that

FIGURE 1.5 (A) The Bowatte Ayurvedic Medicinal Hall in Kandy, Sri Lanka. (B) Many of the traditional medicines can be bought from here in bottled or packet form and are also labelled in English for purchase by local consumers and tourists.

are deemed incurable, locals do always obtain advice from the traditional medicinal practitioners themselves. Despite trends of modernization and urbanization, Sri Lankans still believe in the efficacy of the traditional medicinal system of the country and often will travel significant distances to consult the opinion of traditional medicinal practitioners on matters of complicated disease conditions. Even Sri Lankans living overseas appear to have this confidence. Indeed, many who live abroad (especially elderly Sri Lankans) return to the country just to obtain treatment for diseases that do not seem to have any remedies through 'Western' medicine.

In comparison with traditional Indian or Chinese herbal medicinal systems, the Sri Lankan traditional medicinal system has more unexplored territories that are open for research, since there are few systematic scientific studies on the traditional remedies. Nevertheless, the holistic approach, which is commonly observed in other traditional medicinal systems, is found in the Sri Lankan traditional medicinal system as well (De Zoysa et al. 2017). In Sri Lankan Ayurveda, although the ailment may affect only one part of the body, the treatment is holistic. Additionally, the traditional medicines are not supposed to fight against the ailment; instead, they are supposed to boost the body's immune system to fight against the disorder.

Although several thousand years of clinical evidence exists on the efficacy of certain herbal remedies, the amount of modern scientific research performed on the Sri Lankan traditional medicinal system is considerably less. In addition, these indigenous medicines are yet to reach the overseas consumer market, and quality control measures are yet to be developed across all herbal formulations and implemented to meet the regulatory standards involved in exporting (Karalliadde and Gawarammana 2008). At the same time, in the case of an increased demand due to the popularity of the Sri Lankan herbal remedies, there may be extra pressure placed on natural habitats for larger scale cultivation given the country's lack of available arable land. Thus, prior to using this herb for commercial means, systematic cultivation is needed to ensure the sustainable utilization and conservation, as well as official means of legal and diplomatic protection from exploitation of the species and its bioactive constituents (Waisundara and Watawana 2014a; Karalliadde and Gawarammana 2008).

Despite the many voids and gaps in the science behind the traditional medicinal system of Sri Lanka, the overall medicinal value of plants is well recognized today because of the presence of many natural bioactive substances that are capable of supporting disease prevention or delaying onset of certain diseases. From a scientific perspective, quite possibly owing to the existence of a variety of bioactive compounds in the flora and fauna, the traditional medicinal system of Sri Lanka appears to be robust with a consistently held belief that herbal combinations of botanical and non-botanical remedies may treat any type of disease. Overall, the traditional medicinal system has received many accolades over many decades as an effective therapeutic means of remedying and sometimes preventing many disease conditions. Some of the more important traditional plants, herbs, and medicines are presented in the following chapters. It must be clarified in this instance that this book does not provide a comprehensive compendium of Sri Lankan plants, but only selected ones that are comparatively given lesser attention but which are nonetheless important to the traditional medicinal pharmacopoeia of the country.

REFERENCES

De Zoysa, H.K.S., P.N. Herath, R. Cooper and V.Y. Waisundara. 2017. Paspanguwa herbal formula, a traditional medicine of Sri Lanka: A critical review. *Journal of Complementary Medicine and Alternative Healthcare* 3(2):555609. doi:10.19080/JCMAH.2017.03.555609.

Ediriweera, E.R.H.S.S. and W.D. Ratnasooriya. 2009. A review on herbs used in treatment of diabetes mellitus by Sri Lankan Ayurvedic and traditional physicians. *AYU* 30:373–391.

Karalliadde, L. and I.B. Gawarammana. 2008. *Herbal Medicines: A Guide to Its Safer Use.* London, UK: Hammersmith Press.

Mittermeier, R., N. Myers and C. Mittermeier. 2000. *Hotspots: Earth's Biologically Richest and Most Endangered Terrestrial Ecoregions.* Arlington, VA: Conservation International.

Siriweera, W.I. 1994. *A Study of the Economic History of Pre Modern Sri Lanka.* New Delhi, India: Vikas Publishing House.

Waisundara, V.Y. and M.I. Watawana. 2014a. The classification of Sri Lankan medicinal herbs: An extensive comparison of the antioxidant activities. *Journal of Traditional and Complementary Medicine* 4:196–202.

Waisundara, V.Y. and M.I. Watawana. 2014b. Evaluation of the antioxidant activity and additive effects of traditional medicinal herbs from Sri Lanka. *Australian Journal of Herbal Medicine* 26(1):22–28.

Waisundara, V.Y., M.I. Watawana and N. Jayawardena. 2014. *Salacia reticulata* and its role in diabetes management. *Food and Beverage Asia* 62–64.

Weragoda, P. 1980. The traditional system of medicine in Sri Lanka. *Journal of Ethnopharmacology* 2(1):71–73.

2 Select Examples of Sri Lankan Medicinal Herbs

COCCINIA GRANDIS

Coccinia grandis (L.) Voigt grows abundantly on the Indian sub-continent. Taxonomic information on this plant is provided in Table 2.1. It is known as scarlet gourd, baby watermelon, ivy gourd, or little gourd in English, Tindora in Hindi, Kowakka, Kobowakka, or Kem-wel in Sinhala, Kovai or Kwai in Tamil, and dondakaya in Telugu (Attanayake et al. 2016; Chopra et al. 2018). The fruit of *C. grandis* is very popular in many of the cuisines of South Asia. In Andhra Pradesh of India, deep-fried fruits and cashew nuts or peanuts, sprinkled with some spice mix, is a very popular dish (Chopra et al. 2018). Sometimes, the immature fruits are cooked into a curry and consumed in various parts of the Indian sub-continent as well, whereas the young leaves and long slender stem tops of the plant are cooked and eaten as a pot herb or added to porridge (Waisundara et al. 2015). The young and tender green fruits are sometimes added into raw salads as well.

C. grandis is a very popular anti-diabetic herb in most of the traditional medicinal systems of the South Asian region. In comparison with other anti-diabetic herbs such as *Costus speciosus*, *C. grandis* is mostly administered when more advanced symptoms of the disease appear, such as weight loss, appetite loss, delay in wound healing, or loss of vision. In Sri Lanka, while *C. speciosus* is administered mostly in salad form, *C. grandis* is also consumed as a porridge or as an herbal drink. The flavors of the two plants are very different. *C. speciosus* is tangy whereas *C. grandis* is relatively bland and has a grassy note in salad or porridge form. This plant is also used as a garden ornament in many countries. Some countries in Asia such as Thailand prepare traditional tonic-like drinks for medicinal purposes using the leaves of this plant (Trease and Evans 1978). *C. grandis* leaves are the most popular part of the plant, which is consumed for anti-diabetic purposes in Sri Lanka. This could be considered as a justifiable practice from a biochemical and environmental perspective, because as expected, due to exposure to sunlight and radiation the leaf contains the highest amount of antioxidant compounds, which are of more anti-diabetic importance than the bark and the root (Waisundara and Watawana 2014; Chopra et al. 1956; Marles and Farnworth 1995).

According to a cross-sectional survey carried out in Sri Lanka using an interviewer-based questionnaire, the leaves of *C. grandis* are used as a complementary and alternative medicine by patients with diabetes mellitus (Medagama et al. 2014). *C. grandis* leaves are consumed in salad form in Sri Lanka. They are ground up with a mortar and pestle or a grindstone, added to coconut, and eaten as

TABLE 2.1
Taxonomic Information of *C. grandis*

Kingdom	Plantae
Sub-kingdom	Viridiplantae
Division	Tracheophyta
Sub-division	Spermatophytina
Class	Magnoliopsida
Order	Cucurbitales
Family	Cucurbitaceae
Genus	*Coccinia*
Species	*grandis*

a dish together with rice and curry (Figure 2.1). In a similar manner to *C. speciosus* described in the subsequent section, the recommended amount of consumption of *C. grandis* in this salad form is one heaped tablespoon. It is generally recommended to consume *C. grandis* salad once per day every 2–3 days per week. The hypoglycaemic effects of *C. grandis* can be fast acting and excessive consumption could lead to extremely low blood glucose levels.

FIGURE 2.1 *C. grandis* salad made from leaves and ground together with coconut – a dish consumed in Sri Lanka with rice and curry.

Origins, Morphology, and Growth

Costus grandis belongs to the family Cucurbitaceae. It is native to East Africa, but it has spread widely into tropical Asian countries (Arunvanan et al. 2013). The plant has become naturalized in these Asian countries, mostly because it thrives well in warm, humid, tropical regions (Waisundara et al. 2015).

In Sri Lanka, *C. grandis* is widely distributed in the southern, western, and north central regions of the country although it can grow in other parts including the hilly areas (Attanayake et al. 2016). The plant is a fast-growing perennial and herbaceous climber, which grows several meters tall, spreading over lands that readily cover shrubs and small trees (Figure 2.2a) (Waisundara et al. 2015). The shape of the leaves varies from heart to pentagon, and the leaves are arranged alternately along the stem (Figure 2.2b). The upper surface of the leaf is hairless, whereas the lower surface is hairy (Ediriweera and Ratnasooriya 2009). The flowers of *C. grandis* are large, white, and star-shaped. The fruits are smooth and green, and when ripe, they turn bright red with an ovoid to ellipsoid shape. There are 3–8 glands on the blade near the leaf stalk (Waisundara et al. 2015). The plant has simple tendrils with an extensive tuberous root system. It grows in any soil but prefers a sunny sheltered position in a humus-rich open soil (Waisundara et al. 2015). The plant requires plenty of water during its growing season (Modak et al. 2007). Since the fruits of *C. grandis* are edible, several species of birds enable seed dispersion.

Traditional Medicinal Applications

The medicinal uses of *C. grandis* in the Sri Lankan pharmacopoeia can be traced to an ancient period, where the juice of the roots and leaves were used in the treatment of diabetes, gonorrhea, and constipation (Vaishnav et al. 2001; Rao et al. 2003). The traditional method of administration of the plant was in decoction form with 120 g of fresh leaves or the entire creeper given twice a day (Ediriweera and Ratnasooriya 2009). Most of the diabetic patients in Sri Lanka nowadays prefer to consume the plant in salad form due to ease of preparation. It is generally taken without the advice of a traditional medicinal practitioner. Most of the locals are familiar with the usage of the plant and easy identification and use of the species as an anti-diabetic remedy.

FIGURE 2.2 (a) *C. grandis* creeper and (b) its mature leaf, which is typically used for medicinal purposes.

When consumed in porridge form, ground leaves are added to rice porridge and a combination of several other anti-diabetic herbs such as *Costus speciosus* can also be added to the same mixture. The leaves are never consumed in extract or dried form. Essentially, *C. grandis* plant use in culinary purposes has become the natural source of herbal medicine among Sri Lankans. Other parts of the herb (including the leaves, stems, and roots) have traditional use for the treatment of bronchitis, jaundice, burns, skin eruptions, fever, insect bites, edema, hypertension, stomach pain, dysentery, scabies, allergy, and eye infections (Wasantwisut and Viriyapanich 2003; Abbasi et al. 2009; Sivaraj et al. 2011).

Bioactive Compounds in *C. grandis* and Their Therapeutic Effects

Many secondary metabolites have been identified by High Performance Liquid Chromatography-Mass Spectrometry (HPLC-MS) and Nuclear Magnetic Resonance (NMR) techniques from *C. grandis* (Tamilselvan et al. 2011). The phytochemicals of this plant include saponins, flavonoids, glycosides, xyloglucan, taraxerol, carotenoids, and cryptoxanthin (Ng et al. 2000). Hossain et al. (2014) showed that the ethanol extract of the aerial parts of *C. grandis* contained alkaloids, reducing sugars, and saponins. The saponins, in combination with other phytochemicals, exhibit anti-diabetic activity (Ng et al. 2000). The flavonoids also exhibit antioxidant activity that may also contribute toward anti-diabetic properties. The flavonoids exhibit α-amylase inhibitory activity (Ng et al. 2000; Tamilselvan et al. 2011).

It has been found that *C. grandis* stimulates gluconeogenesis or inhibits glycogenolysis in diabetic rat liver (Munasinghe et al. 2011). Animal studies show that compounds in this plant inhibit the enzyme glucose-6-phosphatase, which is one of the key liver enzymes involved in regulating glucose metabolism (Shibib et al. 1993; Munasinghe et al. 2011). Attanayake et al. (2015) investigated the long-term effects of the aqueous leaf extract of *C. grandis* in streptozotocin-induced diabetic Wistar rats. It was observed that the leaf extract was able to increase the biosynthesis of insulin most likely through β-cell regeneration in the rats. Rahman et al. (2015) investigated the antidiabetic activities of methanolic extract of leaves, fruits, root, and aerial part of *C. grandis* in alloxan-induced diabetic mice. This study indicated that these extracts at dosages of 150, 300, and 450 mg/kg reduced the blood glucose level after 8 hours from the initial dosing (Rahman et al. 2015). Meenatchi et al. (2017) successfully demonstrated *in vitro* the antioxidant and antiglycation potential and insulinotrophic properties of *C. grandis* unripe whole fruits in RINm5F cells. From this study, the authors recommended that the plant might be a promising source for development of natural antiglycaemic agents and novel insulin secretagogues.

Khondhare and Lade (2017) evaluated the phytochemical profile, aldose reductase inhibitory, and antioxidant activities of *C. grandis* fruit extract. In their study, the fruits were extracted with the following solvents: petroleum ether, dichloromethane, acetone, methanol, and water. The methanol extract exhibited the most potent *in vitro* aldose reductase inhibitory activity; however, all extracts exhibited some aldose reductase as well as antioxidant activity (Khondhare and Lade 2017).

Chopra et al. (2018) investigated the antihyperlipidemic activity of ethanolic, methanolic, and chloroform extracts of fruit of *C. grandis* in albino Wister rats. These results were promising: the ethanolic fraction had the highest ability to reduce the serum lipids in comparison with other solvent extracts. The anti-obesity property of *C. grandis* has been claimed but remains to be scientifically fully investigated. Bunkrongcheap et al. (2014) looked at the effects of *C. grandis* leaf, stem, and root on adipocyte differentiation by employing a cell culture model – 3T3-L1 pre-adipocytes upon induction with a mixture of insulin, 3-isobutyl-1-methylxanthine, and dexamethasone as an anti-adipogenesis assay. The root extract exhibited an anti-adipogenic effect (Bunkrongcheap et al. 2014). It significantly reduced intracellular fat accumulation during the early stages of adipocyte differentiation (Bunkrongcheap et al. 2014). It was observed that the inhibitory effect was mediated by at least down-regulating the expression of Peroxisome Proliferator-Activated Receptor-γ (PPARγ) – the key transcription factor of adipogenesis in pre-adipocytes during their early differentiation processes (Bunkrongcheap et al. 2014).

Hasan and Sikdar (2016) evaluated the antibacterial, antifungal, cytotoxic, and pesticidal activities of *C. grandis* roots extract. Antibacterial activity was observed against the tested Gram-positive (*Bacillus subtilis*, *Sarcina lutea*, and *Staphylococcus aureus*) and Gram-negative (*Salmonella typhi* and *Shigella dysenteriae*) bacteria. In antifungal screening, the extract showed moderate antifungal activities against the tested fungi (*Candida albicans* and *Colletotrichum falcatum*) (Hasan and Sikdar 2016). Das et al. (2015) successfully demonstrated the antiproteolytic activity against *L. donovani* serine protease along with its leishmanicidal activity. Laboni et al. (2017) revealed that solvent extracts of *Costus grandis* leaves exhibited moderate antimicrobial activity against some Gram-positive and Gram-negative microorganisms.

Concluding Remarks

C. grandis contains many bioactive compounds, which have an impact on the prevention of many diseases, including its primary condition of application – diabetes. The plant has been successfully used in the Sri Lankan pharmacopoeia for the prevention and treatment of diabetes and its associated complications. Despite the lack of scientifically and systematically studied evidence on its efficacy and toxicological effects, as a traditional medicine, it has always been effective against combating diseases of a complicated nature while the remedial effects have always been holistic, and side-effects have been minimal. As with *C. speciosus*, *C. grandis* is not regulated for use in Sri Lanka; once more the usage is on a personal rather than a commercial basis. Some commercial extracts are available for purchase, but not in Sri Lanka, since the fresh plant is so widely available.

COSTUS SPECIOSUS

Costus speciosus (Koen ex. Retz.) Sm. Costaceae (Family), is widely used in the Sri Lankan Ayurvedic medicinal system primarily for anti-diabetic treatments. It is impossible to disregard and ignore the importance of *C. speciosus* as a key Sri Lankan herbal anti-diabetic treatment. The plant is known by various vernacular terms: crepe ginger or spiral flag in English; Kemuka, Kushta, or Kashmira in Sanskrit; Keukand or Keu in Hindi and Bengali; Chenhalya Koshta in Telengu; and

Thebu in Sinhalese (Waisundara et al. 2015). In Sri Lanka *C. speciosus* is very often grown in home gardens and its leaves are traditionally consumed as a green leaf salad. Different parts of the plant such as the rhizomes, stem, flowers, and roots are applied in Ayurvedic therapy and have been used in the local traditional medicinal system for a long time. The reason behind the popularity of the herb in modern times, other than its medicinal properties, is that it does not carry a repulsive flavor and administration of the plant does not require extensive preparation methods. The plant can be easily grown and does not require special means of nurturing or harvesting and has a very high growth rate. The exact rationale behind the usage and incorporation of *C. speciosus* as a traditional medicinal herb in the Sri Lankan Ayurvedic pharmacopoeia cannot be traced, although its usage for diabetes can be traced almost back 4–5 generations from the present day. In keeping with many anti-diabetic herbs in Sri Lanka, it is possible that traditional medicinal practitioners would have come across this herb through trial and error, including the establishment of a typical recommended dosage.

Origins, Morphology, and Growth

C. speciosus is an ornamental, rhizomatous, and perennial plant belonging to the family Costaceae that consists of about 52 genera and more than 1300 species (Pawar and Pawar 2012; Waisundara et al. 2015). Within the Zingiberales, the family costaceae is easily recognized and distinguished from other families by its well-developed and occasionally branched aerial shoots, which possess characteristic monistichous (one-sided) spiral phyllotaxy (Pawar and Pawar 2012). The taxonomic classification of the *C. speciosus* plant is described in Table 2.2.

C. speciosus is native to South East Asia although it is currently more abundantly found and used for medicinal purposes in India, Sri Lanka, Indonesia, and Malaysia (Arunvanan et al. 2013). Within India, *C. speciosus* grows in abundance throughout the foothills of the Himalayas from the areas of Himachal Pradesh to Assam, the Vindhya Satpura hills in Central India, Eastern Ghats of Andhra Pradesh and Western Ghats of Maharashtra, Karnataka, Tamil Nadu, and Kerala

TABLE 2.2
Taxonomic Classification of *C. speciosus*

Kingdom	Plantae
Sub-kingdom	Tracheobinota
Super-division	Spermatophyta
Division	Mangoliophyta
Class	Liliopsida
Sub-class	Zingiberidae
Order	Zingiberales
Family	Coastaceae
Genus	*Costus*
Species	*speciosus*

(Sarin et al. 1974). The plant nevertheless has been naturalized in some tropical areas such as Hawaii. *C. speciosus* is so common and abundant in certain parts of Sri Lanka that it may even be considered a weed.

Morphologically, *C. speciosus* is an erect, succulent herb up to 2.7 m high, arising from a horizontal rhizome (Arunvanan et al. 2013). Red stems emerge from the rhizome, which bears the large and soft, variegated leaves (Waisundara et al. 2015). The leaves are typically elliptical to oblong or oblong-lancoelate, which are spirally arranged and have a silky underside (Figure 2.3). They are large, approximately 1.5″ in diameter with thick, cone-like terminal spikes, bright red bracts, and lip-white with a yellowish centre, crisped, concave disk, and a tuft of hair at the base (Modak et al. 2007; Waisundara et al. 2015). Some varieties of *C. speciosus* contain flowers and bracts that appear with compact cones, while others are shaped like a pineapple or soft crepe coming out of green cones (Trease and Evans 1978).

C. speciosus is mainly cultivated in areas that receive significant amounts of rainfall and it grows well on fertile moist soil or clay loam soil in shady areas (Modak et al. 2007). A level of high humidity and low temperature provide the best conditions for cultivation of this plant. In Sri Lanka, it can be found growing well in hilly areas. The multiplication rate, percentage of seed germination, and seed viability are low in this plant. Thus, for it to be grown in mass as an agricultural crop, various vegetative methods such as using rhizome pieces, division of culms, and stem cutting are typically used. The most natural method of propagation of *C. speciosus* is by seeds dispersed by birds.

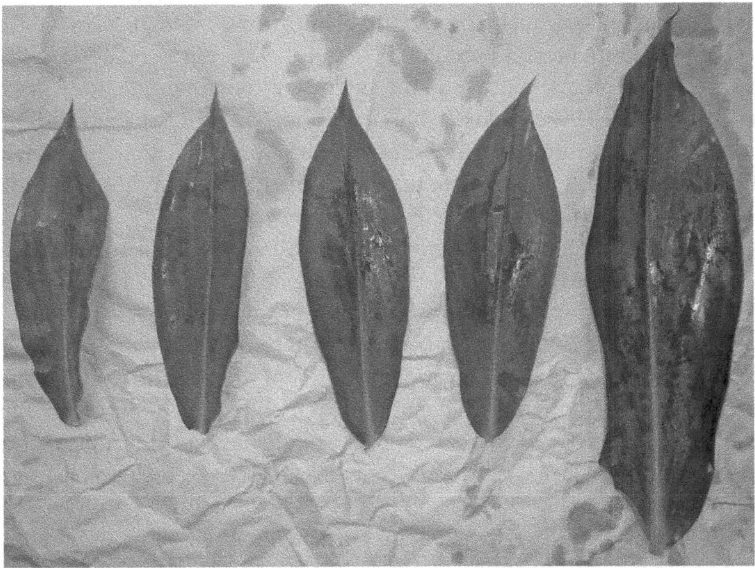

FIGURE 2.3 Leaves of *C. speciosus* can be readily obtained from the marketplace in Sri Lanka. The shorter ones to the left are young and not typically used for medicinal purposes but are often consumed in salad form with rice and curry; the longer and more mature leaf on the right is used for medicinal purposes in addition to being consumed as a culinary dish.

Delays have been observed in the plant's growth when employing methods that are typically used for large-scale production in preservation and commercial cultivation. Therefore, many biotechnological approaches and initiatives such as tissue culture and germplasm preservation are currently being evaluated for possible mass production of *C. speciosus* on the Indian sub-continent (Neelesh et al. 2010).

It is easy to recognize the variety of *C. speciosus* that is typically used for medicinal purposes, and therefore, no mis-identifications of the plant have been reported to date. Other varieties of the plant that are available in Sri Lanka have distinguishable marks on their leaves (mostly a red outline on the border of the leaf). Thus, local people easily recognize a *C. speciosus* plant with leaves *without* any markings for consumption.

Traditional Medicinal Applications

Both the leaves and roots of the plant are used in all the traditional medicinal systems of the Indian sub-continent. The extracts obtained from both these plant parts are considered to have a bitter, astringent, acrid, and cooling effect upon consumption, while it is known for therapeutic effects as an aphrodisiac, purgative, anthelmintic, depurative, febrifuge, expectorant, and tonic (James et al. 2009). Apart from anti-diabetic effects, *C. speciosus* is also useful in treating burning sensations, leprosy, worm infections, skin diseases, fever, asthma, bronchitis, inflammations, and anaemia. The rhizome in particular is widely administered in the traditional medicinal systems of the region as a remedy for pneumonia, constipation, fever, asthma, bronchitis, inflammation, anaemia, rheumatism, dropsy, cough, urinary diseases, and jaundice (Moosmann and Behl 2002).

In contrast with other countries in the Indian sub-continent, *C. speciosus* leaves are the most popular form used for medicinal purposes in Sri Lanka. The leaves are available for harvesting any time of the year. As a tropical country, Sri Lanka has the advantage of reaping most of the medicinal herbs at any time, and these are typically ever-greens like *C. speciosus*. The *C. speciosus* leaves are generally cut or torn into pieces and ground in a mortar and pestle or a grindstone together with coconut (Figure 2.4). A small amount of lime is also added to contribute a sweet and sour flavor, and typically this mixture is consumed as a salad together with rice and curry. Due to its almost spontaneous hypoglycaemic effects, it is generally recommended to add one heaped tablespoon of the plant mix to the salad and consume once every 2 or 3 days. There have been no recorded reports on the toxicity of the herb (especially those leading to hospitalization or mortality), other than occasional significant drops in blood sugar levels upon excessive consumption of the *C. speciosus* salad.

The anti-diabetic effects of *C. speciosus* have made the plant ever more popular among modern-day diabetics in Sri Lanka given the rising incidence of the disease. The prevalence rate of diabetes in the country (5.02%) was found to be associated with age-adjusted impaired glucose intolerance (5.27%) in the rural communities in Sri Lanka (Katulanda et al. 2012). A recent study conducted in Sri Lanka demonstrated that the usage of herbal medicines is 76% among a group

FIGURE 2.4 *C. speciosus* salad being ground and prepared for consumption with rice and curry in a mortar and pestle.

of 252 type 2 diabetic patients investigated (Medagama et al. 2014). These patients were also observed to take one or more oral hypoglycaemic agents. Among them, 47% have consumed *C. speciosus* leaf as a salad in their main meals. This popularity was found to be mostly due to *C. speciosus* leaves being more abundant and cheaper when obtained from the local marketplaces than commercially prepared products. Also, the ease of preparation is another factor that has led to the popularity of this herb in Sri Lanka. At present, many scientific studies have been carried out or are ongoing on *C. speciosus* to exploit its beneficial effects for commercial purposes. Many of these studies are in *in vitro* or *in vivo* phases and have not, as yet, progressed into human clinical trials (Waisundara et al. 2015). Nevertheless, despite the relative lack of modern scientific evidence, *C. speciosus* is consumed in Sri Lanka for general health and wellness purposes owing to its historical reputation as a valuable medicinal herb.

Bioactive Compounds in *C. speciosus* and Their Therapeutic Effects

The major secondary metabolites of this plant that impart anti-diabetic effects are alkaloids, flavonoids, glycosides, phenols, sterols, and sesquiterpenes. HPLC-MS has revealed the presence of significant amounts of saponins in the herb, and the major saponin that has been isolated from the seeds of this plant is diosgenin (Figure 2.5a) (Bavarva and Narasimhacharya 2008). Diosgenin is a steroid and is the major constituent isolated in the rhizome as well (Bavarva and Narasimhacharya 2008). The maximum quantity of diosgenin reported in the stem is 0.65%, in the leaves 0.37%, and in the flower 1.21% (Bavarva and Narasimhacharya 2008). Selim and Al Jauoni (2016)

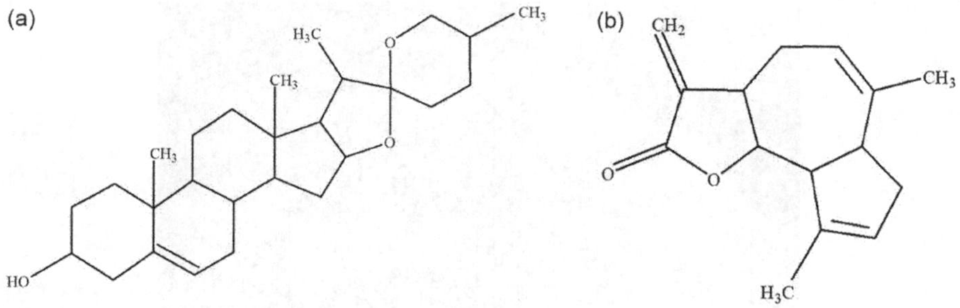

(a) (b)

FIGURE 2.5 Two of the major bioactive compounds in *C. speciosus*: (a) Diosgenin and (b) Eremanthin.

conducted a study to evaluate the *in vitro* anti-inflammatory, antioxidant, and anti-angiogenic activities of diosgenin isolated from *C. speciosus.* The study outcomes suggested that diosgenin possesses anticancer, apoptotic, and inhibitory effects on cell proliferation. Eremanthin is the second most potent and abundant anti-diabetic bioactive compound isolated from *C. speciosus* (Figure 2.5b). This compound has been tested in streptozotocin-induced diabetic Wistar rats and has resulted in reduced plasma glucose levels (Eliza et al. 2009). Other than these two compounds, dioscin, gracillin, β-sitosterol, and β-D-glucoside were isolated in the rhizome and identified using HPLC-MS and NMR.

Many investigations have demonstrated the reduction of hyperglycaemic effects using *C. speciosus* rhizome *in vitro* or *in vivo* (Bavarva and Narasimhacharya 2008; Revathy et al. 2014; Ali Almaghrabi and Afifi 2014). Specifically, the oral administration of the ethanolic extract of *C. speciosus* rhizome on the levels of biochemical parameters and activities of enzymes in alloxan-induced diabetes rats were analyzed. The study showed that the plant extract enhances insulin secretion by the islets of Langerhans, enhances peripheral glucose utilization, and increases serum protein levels. In comparison, much less systematic research work has been performed on the *C. speciosus* leaf extract. In a study by Samarakoon et al. (2013), the antioxidant effects of different solvent-soluble fractions derived from the methanolic extracts of *C. speciosus* leaves were assessed. The ethyl acetate and aqueous fractions possessed the highest antioxidant activity. Perera et al. (2016) demonstrated *in vitro* that the methanol extracts of *C. speciosus* leaves imparted inhibitory activities on α-glucosidase, fructosamine formation, glycation, and glycation-induced protein cross-linking. These findings provide some modern scientific evidence in support of the use of *C. speciosus* leaves for hypoglycaemic effects with an added advantage in slowing down protein glycation.

Regarding anti-bacterial activity of the plant, it has been observed that the rhizome extract is effective against Gram-positive (*Staphylococcus aureus*, *Staphylococcus epidermidis*) and Gram-negative bacteria (*E. coli*, *Pseudomonas aeruginosa*, *Salmonella typhimurium*), in part due to the presence of diosgenin (Vijayalakshmi and Sarada 2008; Ariharan et al. 2012). Nair et al. (2014) investigated the effects of methanol extract of leaves of *C. speciosus* on the growth of human hepatocellular carcinoma (HepG2) cells and they elucidated its possible mechanisms of action.

It was observed that the methanolic extract perturbed cell cycle progression, modulated cell cycles, and regulated signal molecules that were involved in induction of apoptosis in HepG2 cells. Susanti et al. (2018) conducted a study to determine the effect of the aqueous extract of *C. speciosus* on cholesterol blood serum levels of rat-induced propylthiouracil (PTU). It was shown that administration of *C. speciosus* extract at a dosage of 200 mg/kg can reduce the total serum cholesterol equal to the anti-hypercholesterolemic drug simvastatin at 7.2 mg/kg. Kala et al. (2015) investigated the anti-arthritic potential of methanolic extract of rhizome of *C. speciosus* in Wistar rats. The anti-arthritic activity was shown by the prophylactic high-dose test extract (200 mg/kg), which at this dose was as potent as the standard drug indomethacin (10 mg/kg).

Concluding Remarks

C. speciosus is a promising herb with much potential to be developed into an effective anti-diabetic medication with worldwide recognition. From the evidence to date, the plant has also demonstrated pharmacological activities *in vitro* and *in vivo* such as anti-inflammatory, anti-microbial, antioxidant, anti-dyslipidemic, and anti-cancer benefits. It only remains to hope that through additional scientific findings and human clinical trials broader studies for diabetes management using the herb can be explored. Further, the bioactive compounds diosgenin and eremanthin may also be evaluated individually and in combination and might lead toward acceptance in the global market of pharmaceuticals. To date in Sri Lanka, there are no special regulations governing the usage of the herb; the consumption is carried out on an individual basis and is well tolerated by the general population even without the recommendation of an Ayurvedic practitioner. Typically, extracts or dried forms of the plant are rarely sold in the Ayurvedic medicinal halls of Sri Lanka, and there are no regulations or restrictions in place. Overall, *C. speciosus* remains a very important anti-diabetic herb in the Sri Lankan Ayurvedic pharmacopoeia and as mentioned previously, with the rising incidence of diabetes, its popularity will only keep growing in modern times.

PLECTRANTHUS AMBOINICUS (LOUR.) SPRENG

Before going into details of *P. amboinicus* itself, it is important to describe the genus to which it belongs, since the plants under this category have a variety of purposes and interesting characteristics in general. The genus *Plectranthus* L'Hér. (Lamiaceae), also known as spurflowers, belongs to the subfamily Nepetoideae, tribe Ocimeae, and subtribe Plectranthinae, comprising approximately 300 species distributed through the tropical and warm regions of the Old World (Rice et al. 2011). The diversity of *Plectranthus* is an important element of the biodiversity and traditional medicinal systems of Asia as well as Africa. Many *Plectranthus* species are used as herbal medicines in these continents and have the potential for development towards their use in the primary health care system (Gaspar-Marques et al. 2006). In fact, *Plectranthus* is most commonly cited in the literature for its medicinal properties and uses (Lukhoba et al. 2006). Plants belonging to the genus *Plectranthus* are attractive and floriferous, requiring little light (shade tolerant), adaptability to semi-dry conditions, and tolerance to a warm and dry atmosphere (Rice et al. 2011). Furthermore, these plants are robust and easy to grow. It is a useful genus for developing gardens and landscapes, which may need large areas covered with shrubbery within a short duration of time, since this can be easily achieved with *Plectranthus* plants using cuttings (Rice et al. 2011). Shrubby and groundcover *Plectranthus* plants assist with a build-up of leaf mulch that fertilizes the soil, traps moisture, and out-competes weeds. Several bee species, long-proboscid flies, generalist butterflies, and day-flying hawkmoths visit *Plectranthus* flowers for nectar, thus taking part in the ever-important pollination process (Rice et al. 2011).

P. amboinicus is known as 'kapparawalliya' in the vernacular language of Sinhala in Sri Lanka and is grown in many backyards for horticultural purposes rather than for medicinal value. It is also known as country borage, Indian borage, 'karpuravalli' and 'omavalli' in Tamil, and 'patta ajavayin' and 'patharcur' in Hindi (Kaliappan and Viswanathan 2008). A list of vernacular names that have been used across several countries to refer to *P. amboinicus* is shown in Table 2.3. The leaves of the plant are so aromatic that even touching them lightly would leave behind a pungent smell on the hand. The leaf structure of a young potted plant is shown in Figure 2.6. The plant is also used for horticultural purposes in rockeries in South Africa (Van Jaarsveld 2006). This herb is also widely used by indigenous people of tropical rain forests, either in folk medicine or for culinary purposes (Arumugam et al. 2016). The leaves

TABLE 2.3
Vernacular Names of *P. amboinicus* Commonly Used by Locals in Their Respective Countries

Country	Vernacular Name
Barbados	Poor man's pork, broad-leaf thyme
Cambodia	Sak dam ray
China	Da shou xiang
Cuba	Orégano, orégano de Cartagena
Fiji	Rhaivoki, Sage
Germany	Jamaika thymian
Guyana	Thick-leaf thyme, broad-leaf thyme
Indonesia	Torbangun, daun kutjing
Malaysia	Daun bangun-bangun, pokok bangun-bangun
Philippines	Latai, suganda, oregano
Puerto Rico	Puerto Rican oregano brujo, Cuban oregano
South Africa	Sup mint, French thyme, Indian mint
Thailand	Hom duan huu suea, niam huu suea
USA	Indian borage, country borage, Spanish thyme, Mexican mint, French thyme, Indian mint
Vietnam	Can day la
West Indies	French thyme, Spanish thyme, broad-leaf thyme

Source: Arumugam, G. et al., *Molecules*, 21, 369, 2016.

FIGURE 2.6 Leaf structure of a *P. amboinicus* plant, which is grown in the gardens of Sri Lanka most of the time as a pot-herb.

of the plant are often eaten raw by Sri Lankans or used as flavoring agents, incorporated as ingredients in the preparation of traditional food. The chopped leaves are also used as a substitute for sage (*Salvia officinalis*) in meat stuffing as shown in historical evidence obtained from *ola leaf* records. The essential oil is obtained from the leaves and stem explants of *P. amboinicus* where it was shown to contain at least 76 volatile constituents (Arumugam et al. 2016). In the essential oil, the two major phenolic compounds are carvacrol and thymol, which have been used for various culinary properties (Arumugam et al. 2016).

Origins, Morphology, and Growth

The taxonomic classification of *P. amboinicus* is shown in Table 2.4. It is a perennial with a 3–10-year lifespan, and is distributed in Tropical Africa, Asia, and Australia (Chang et al. 2010). Other than for medicinal purposes, this plant is used as a food, an additive, and fodder (Chang et al. 2010). The *P. amboinicus* plant consists of a hispidly villous or tomentose, fleshy stem about 30–90 cm. Its leaves are simple, broad, ovate, and very thick, studded with hairs; on the lower surface, the glandular hairs are most numerous and give rise to a frosted appearance (Kaliappan and Viswanathan 2008). The taste of the leaf is pleasantly aromatic with an agreeable and refreshing odor. Flowers are shortly pedicelled, pale purplish, and come in dense whorls at distant intervals in a long slender raceme (Kaliappan and Viswanathan 2008). The flowers have a bell-shaped calyx and the throat is smooth inside with two lips, the upper lip being ovate and thin, the lower lip having four narrow teeth (Arumugam et al. 2016). The corolla is pale purplish and five times longer than the calyx, with a short tube, inflated throat, and short lips (Roshan et al. 2010; Khan 2013). Fruit and

TABLE 2.4
Taxonomic Classification of *P. amboinicus*

Kingdom	**Plantae**
Sub-kingdom	Viridiplantae
Division	Tracheophyta
Sub-division	Spermatophytina
Class	Magnoliopsida
Super-order	Asteranae
Order	Lamiales
Family	Lamiaceae
Genus	*Plectranthus* L'Hér.
Species	*Plectranthus amboinicus* (Lour.) Spreng

nuts are smooth, pale brown in color, 0.7 mm long and 0.5 mm wide. *P. amboinicus* rarely flowers and seeds are difficult to collect (Khan 2013).

P. amboinicus is a fast-growing plant usually propagated by stem cuttings. This preferred method of propagation through vegetative means is rarely followed nowadays, though, because it rarely seeds or sets seed (Khan 2013). The herb grows easily in a well-drained, semi-shaded location. It is found to grow well under tropical and subtropical locations (Arumugam et al. 2016). It was also found to adapt well in cooler climates if grown in a pot and brought indoors or moved to a warm, sheltered position during winter. The plant needs little water. *P. amboinicus* grows best in rich, compost soil with neutral pH and high humidity, but if there is excess water in the ground its roots rot (Staples and Kristiansen 1999). It copes well with severe droughts, as it has lots of water stored in its succulent flesh (Arumugam et al. 2016). It also survives well with severe heat and scorching sun, as well as strong shade, but grows best under partial shade (Arumugam et al. 2016).

Traditional Medicinal Applications

In Sri Lanka, extracts of *P. amboinicus* have been used for antibacterial effects and to prevent infections. The same application has been observed in many of the traditional medicinal systems of the Indian sub-continent. A decoction of the leaves has been used for several medicinal purposes in the traditional medicinal system of India, especially against respiratory diseases such as congestion, bronchitis, sore throat, and issues in the digestive tract such as dysentery, diarrhea, and colitis (Bhatt et al. 2013). In certain tribes of India, the plant is used to treat malarial fever, hepatopathy, renal and vesical calculi, cough, chronic asthma, bronchitis, hiccups, and epilepsy (Devi and Periyanayagam 2010). The plants are commonly used in traditional Chinese medicine for the treatment of cough, fever, sore throats, mumps, and mosquito bites (Chiu et al. 2012).

In Africa, an infusion or syrup made from the aromatic leaves of *P. amboinicus* is prescribed for the treatment of coughs (Rabe and Van Staden 1998; Albuquerque 2001). The plant is also an important ingredient in many of the Zulu herbal medicinal formulations, while the leaves are used as a flavoring agent to food (Hutchings et al. 1996).

P. amboinicus is one of the plants traditionally used in the treatment of leishmanial ulcers due to *Leishmania (vlannia) braziliensis* in the endemic area of Bahia, Brazil according to França et al. (1996). According to this report, nearly 33% of the population in this region preferred and cited using this plant for this specific ailment. *P. amboinicus* is one of the most frequently cited species for the treatment of burns, wounds, sores, insect bites, and allergies in the Democratic Republic of Congo (Lukhoba et al. 2006).

Bioactive Compounds in *P. amboinicus* and Their Therapeutic Effects

El-Hawary et al. (2012), isolated 3-methoxy genkwanin, crisimaritin, *p*-coumaric acid, caffeic acid, taxifolin, rosmarinic acid, apigenin, and 5-O-methyl-luteolin from the ethyl acetate fraction of *P. amboinicus*. The bioactivities of these compounds are well recognized, and thus, it was assumed in this study that the therapeutic properties of the herb are purportedly due to the existence of these compounds (El-Hawary et al. 2012). A GC-MS screening of the alcoholic extract of leaf powder of *P. amboinicus* conducted by Uma et al. (2014) revealed the presence of several other compounds, which are shown in Table 2.5. Their known therapeutic properties are indicated in the table as well.

TABLE 2.5
Phytoconstituents Identified by Uma et al. (2014) in the Alcoholic Leaf Extract of *P. amboinicus* Using GC-MS Analysis

Phytoconstituent	Demonstrated Biological Activities
3,7,11,15-Tetramethyl-2-hexadecen-1-ol	Antimicrobial, anti-inflammatory
9,12-Octadecadienoic acid (Z, Z)	Hypocholesterolemic, anti-arthritic, hepatoprotective, anti-androgenic, anti-helminthic, anti-histaminic, anti-eczemic
Eugenol	Analgesic, antibacterial, anti-inflammatory, antioxidant
n-Hexadecanoic acid	Antioxidant, hypocholesterolemic, nematicide, anti-androgenic, haemolytic inhibitor
γ-Tocopherol	Antioxidant, vasodilator, anti-tumor, analgesic, hepatoprotective, anti-diabetic
Thymol	Antimicrobial, anti-inflammatory, antiseptic, antioxidant, anti-cancer
Tetradecanoic acid	Antioxidant, anti-cancer, anti-helminthic, hypocholesterolemic
Phytol	Anti-cancer, antimicrobial, antioxidant
Squalene	Anti-cancer, antimicrobial, antioxidant
Oleic acid	Anti-inflammatory, anti-androgenic, anti-cancer, hypocholesterolemic
Borneol	Anti-pyretic, anti-inflammatory, insect repellent, nematicidal, sedative, analgesic, antibacterial
1-Pentanamine	Antibacterial
3-Methyl-4-isopropyl phenol	Antioxidant, antibacterial, analgesic
9,12,15-Octadecatrien-1-ol (Z,Z,Z)	Antimicrobial
Caryophyllene oxide	Anti-cancer, anti-inflammatory, analgesic, anti-bacterial
Eudesma-4(14),11-diene	Anti-cancer, sedative, anti-inflammatory, anti-fungal
Durohydroquinone	Antimicrobial, antioxidant, sedative

The antioxidant potential of the aqueous and ethanolic extracts of leaves of *P. amboinicus* were investigated by Patel et al. (2010). The preliminary phytochemical screening in this instance showed the presence of alkaloids, carbohydrates, glycosides, proteins, amino acids, flavonoids, quinine, tannins, phenolic compounds, and terpenoids. The antioxidant potential and reducing power of both extracts appear to have increased with increasing concentration of extract in this study. In the study by Bhatt et al. (2013), the stem of *P. amboinicus* was found to be an antioxidant-rich fraction as evaluated by *in vitro* models such as DPPH free radical scavenging activity assay, reducing power assay, superoxide anion radical scavenging, and total antioxidant capacity assays. The extract also exhibited antiplatelet aggregation ability, antibacterial activity, and antiproliferative effects against the cancer cell lines of Caco-2, HCT-15, and MCF-7 (Bhatt et al. 2013). Bhatt and Negi (2012) also demonstrated the antioxidant effects of the ethyl acetate fraction of the leaves of *P. amboinicus* using *in vitro* antioxidant assays.

P. amboinicus contains essential oils, flavonoids, and terpenes that possess inhibitory effects against Gram-positive and Gram-negative bacteria (Gurgel et al. 2009; Bhatt and Negi 2012). The study by Gurgel et al. (2009) describes some *in vitro* experiments with the hydroalcoholic extract of leaves from *P. amboinicus* in several methicillin resistant *Staphylococcus aureus* (MRSA) strains. Outcomes of this study corroborated the use of *P. amboinicus* in many of the traditional medicinal systems in the Indian sub-continent for the treatment of infections caused by *S. aureus*. In the study by Vijayakumar et al. (2015), zinc oxide nanoparticles were biologically synthesized using the leaf extract of *P. amboinicus*. It was observed that the growth of MRSA biofilms (MRSA ATCC 33591) was inhibited by the nanoparticles at the concentration of 8–10 μg/mL. Bhatt and Negi (2012) demonstrated that the ethyl acetate extract of the leaves of *P. amboinicus* had antibacterial effects against several foodborne pathogens, namely, *S. aureus*, *Bacillus cereus*, *E. coli*, and *Yersinia enterocolitica*.

Chang et al. (2010) investigated the therapeutic efficacy of *P. amboinicus* aqueous leaf extract in treating rheumatoid arthritis using a collagen-induced arthritis animal model. It was observed in this study that *P. amboinicus* significantly inhibited the footpad swelling and arthritic symptoms in collagen-induced arthritic rats. In the *in vitro* and *in vivo* study by Chiu et al. (2012) the analgesic and anti-inflammatory properties of the aqueous extract from *Plectranthus amboinicus* were investigated. The extract inhibited pain induced by acetic acid and formalin and the inflammation induced by carrageenan in experimental mice. The anti-inflammatory effect of the extract was related in this study to modulating antioxidant enzyme activities in the liver and decreasing the malondialdehyde level, along with the production of tumor necrosis factor-α, and cyclooxygenase-2 in edema-paw tissue in mice. Analysis of the *in vitro* portion of this study showed that *P. amboinicus* aqueous extract inhibited the pro-inflammatory mediators in RAW 264.7 cells stimulated with lipopolysaccharide. Overall, Chiu et al. (2012) demonstrated that the effects observed in the study provide evidence for traditional medicinal uses of *P. amboinicus* in relieving pain and inflammation. In the study by Devi and Periyanayagam (2010), it was observed that the anti-inflammatory activity of the ethanolic extract of *P. amboinicus* was comparable to that of the standard drug, hydrocortisone.

The study by Shenoy et al. (2012) demonstrated the hepatoprotective effect of *P. amboinicus* ethanolic extract in experimental rats with paracetamol-induced hepatotoxicity. The extract improved the histopathology of the liver in the rats at the dosages of 600 and 900 mg/kg. The anti-diabetic and anti-hyperlipidemic effects of *P. amboinicus* ethanolic extract was observed in alloxan-induced diabetic Wistar rats (Vishwanathaswamy et al. 2011). The extract almost normalized the pancreatic structure in the rats at the end of the study period of this particular investigation. The diuretic effect was observed in various solvent extracts of *P. amboinicus* leaves in male albino rats in the study by Patel et al. (2010).

Essential oil of *P. amboinicus* was studied for its larvicidal potential against the malarial vector mosquito *Anopheles stephensi* by Senthilkumar and Venkatesalu (2010). The results of this study showed that the essential oil of *P. amboinicus* is one of the most inexpensive and eco-friendly sources of natural mosquito larvicidal agents to control and reduce the population of malarial vector mosquitoes. Antimicrobial and anti-inflammatory activity of extracted essential oil of *P. amboinicus* was tested in experimental animal models by Manjamalai, Alexander, and Grace (2012). The inflammation was found to reduce to normal levels in essential oil-treated rats in a similar manner to the standard drug, diclofenac.

Concluding Remarks

Locally, the medicinal value of *P. amboinicus* has not gained full recognition, mostly owing to the cultivation of the plant for horticultural purposes. In fact, it is quite normal for many Sri Lankans to purchase potted versions of the plant for decorative purposes, rather than for medicinal effects. This is one distinctive feature that sets *P. amboinicus* apart from cultivation of other traditional medicinal herbs of Sri Lanka such as *C. grandis* and *C. speciosus*, where it is primarily grown by locals for landscaping. The plant is rarely consumed in salad form or as an herbal porridge, although it is used as a flavoring agent in a similar manner to oregano for various culinary preparations. The essential oil of this plant is used for various pharmaceutical purposes, especially as a cough syrup, but these products are not as popular as those consisting of *Phaseolus adenanthus* ('valmi' or liquorice root), *Coscinium fenestratum* ('veniwel gata'), or *Oldenlandia corymbosa* ('pathpadagam') as ingredients that are considered more potent against respiratory disorders. There are few clinical studies and this presents a significant void when it comes to obtaining a holistic picture of the therapeutic effects of *P. amboinicus*, and thus opportunities may exist to discover whether any novel bioactive compounds exist in the plant.

KALANCHOE LACINIATA (L.) DC

Around 50–60 years ago, it was the mischievous habit of school children of Sri Lanka to preserve a leaf of *K. laciniata* within the pages of their school text books to see roots spring out from it after a certain period. Since text books are made of tree pulp, the moisture content triggers shoots of roots to appear from the leaf. The actual method of propagation of the plant is through its leaves as well and does not require soil. Sadly, this connection of observing nature does not occur in modern-day school children of the country, regardless of their residing in urban or rural areas. Thus, identification of *K. laciniata* plants and the memories of playing with its leaves during school time lies within the older generation whose population is declining. *K. laciniata* is a highly overlooked plant in the traditional medicinal pharmacopoeia of Sri Lanka. Its popularity has reduced with time, giving space to 'herbal giants' that have a broader commercial value such as *Centella asiatica* and *Murraya koenigii*. The plant is called 'Akkapāna' in Sinhala, 'Mallakulli' in Tamil, 'Parnabija/Asthibhaksha' in Sanskrit, and Cathedral Bells or Air Plant in English. The leaf structure and plant are shown in Figure 2.7, while the taxonomic classification is shown in Table 2.6. Another interesting fact is that one of the plants from the *Kalanchoe* genus was the first to be sent into space, propelled on a resupply to the Soviet Salyut 1 space station in 1971 (Chernetskyy 2011).

FIGURE 2.7 (a) Aerial leaf structure and (b) shape of the plant of *K. laciniata*.

TABLE 2.6
Taxonomic Information of *K. laciniata*

Kingdom	**Plantae**
Sub-kingdom	Viridiplantae
Division	Tracheophyta
Class	Magnoliopsida
Order	Saxifraganae
Family	Crassulaceae
Genus	*Kalanchoe* Adans
Species	*Kalanchoe laciniata* (L.) DC.

Origins, Morphology, and Growth

K. laciniata plant is commonly found in home gardens in Sri Lanka and sometimes seen as a weed in shady places. It typically grows in the low-country dry zone of Sri Lanka. It is a perennial herb, about one 1 m in height, glabrous and with stout stems. The upright stems are fleshy (i.e. succulent) and hairless. Leaves are bipinnate, very succulent, 7.5–10 cm in length, reaching up to a width of 20 cm; the plant has a petiole-like base 2–4 cm in length. The leaves are also fleshy (i.e. succulent) and oppositely arranged and flattened. These leaves are green or yellowish-green in color, hairless (i.e. glabrous). Inflorescence is 12–25 cm in length. The flowers are generally yellow in color with the pedice being 9–16 mm in length; the corolla tube is 5–6 mm in length. The fruit is follicular, papery, and membranous, 0.8–1 cm in length. The genus *Kalanchoe* consists of about 125 species of tropical, succulent flowering plants (Deb and Dash 2013). These are mostly cultivated as ornamental plants. They are popular for this specific application, because of their ease of propagation, low water requirements, and displaying wide varieties of flower colors typically borne in clusters well above the phylloclades. In the past, the genus *Kalanchoe* was divided into three genera: *Kalanchoe*, *Bryophyllum*, and *Kitchingia*, but today, most botanists recognize it as one genus (Burrows and Tyrl 2001).

Traditional Medicinal Applications

In Sri Lanka, the extract of *K. laciniata* leaves has been traditionally applied to wounds and contusions and used for preventing swelling. The leaf extract is orally administered for kidney stones as well; however, in this instance, there appears to be no clear record of its dosage. *K. laciniata* leaves have been used as an astringent, antiseptic, and counter-irritant against poisonous insect bites in Sri Lanka. This species nevertheless is thought to be poisonous to livestock. Records have shown that excessive ingestion of the leaves causes cardiac poisoning, particularly in grazing animals. Other than in extract form, the leaves are also consumed as a salad.

The *K. laciniata* plant is distributed throughout Africa, southern India, Pakistan, Myanmar, Thailand, China, Java, and Brazil and is a component of the traditional medicinal systems of these regions as well (Manan et al. 2015). However, common to all traditional medicinal systems, the leaves are roasted and applied to relieve inflammation (Manan et al. 2015). In India, it is used as an emollient and to relieve headache and joint pain (Karuppuswamy 2007) and for treating smallpox (Deb and Dash 2013). In Indo-China, the juice of the plant is taken to treat burns and bruises and superficial ulcers (Manan et al. 2015). It is also reported that the Palian tribes in the Sirumalai hills of southern India use the leaf juice externally for joint pain (Karuppuswamy 2007). The plant is also used to cure diabetes, jaundice, and rheumatism in the Mysore and Coorg districts of the Karnataka state of India (Kshirsagar and Saklani 2007).

There are a few ethnobotanical surveys that mention the usage of *K. laciniata* for various medicinal purposes. In the ethnobotanical survey by Bhandary et al. (1995), the Siddis of Uttara Kannada district in Karnataka, India used *K. laciniata* leaves for scabies and leukoderma, while the decoction of the leaves was used for cuts to stop bleeding. The Siddis are descendants of African slaves who were brought to Goa, India during the occupation of the Portuguese. It is believed that the traditional medicinal knowledge brought by this community was passed on to the residents of the area, including the usage of *K. laciniata* (Bhandary et al. 1995). In the ethnobotanical survey by Megersa et al. (2013), fresh or dried root of *K. laciniata*, seed of *Capsicum frutescens*, *Allium sativum*, and leaves of *Croton macrostachyus* are powdered together and given for various cattle ailments including blackleg in Wayu Tuka District, East Welega Zone of Oromia Regional State, West Ethiopia. Sajeev and Saseedharan (1997) conducted an ethnobotanical survey of the Chinnar Wildlife Sanctuary in the state of Kerala in India. The tribes in the sanctuary are Hill Pulayas and Muthuvans, living at 11 settlements, where the Muthuvans are considered as a superior group and do not mingle with the Hill Pulayas. It was observed that both these tribes use the leaf extract of *K. laciniata* on wounds. Birhanu et al. (2015) conducted an ethnobotanical survey in selected areas of Horro Gudurru Woredas, Western Ethiopia. A total of 81 major medicinal plant species belonging to 43 families was documented in this survey, along with details of their local name, family, habit and traditional preparation, and mode of application. *K. laciniata*, known as 'Bosoqqee', was used for wound-healing by the locals. The whole plant and root were used for this purpose.

Bioactive Compounds in *K. laciniata* and Their Therapeutic Properties

The phytochemical screening of *K. laciniata* was carried out by Manan et al. (2015), with aqueous-methanol and n-hexane extracts. The n-hexane extract consisted of tannins and terpenoids, while the aqueous methanolic extract contained saponins, tannins, terpenoids, flavonoids, glycosides, and anthraquinones. Anderson et al. (1983) reported three toxic bufadienolides, one characterized as hellibrigenin-3-acetate from the plant (Figure 2.8), but the bioactivity of this compound has not been elucidated to date. Bufadienolides are a group of polyhydroxy C-24 steroids and their glycosides that have been identified to be present throughout the *Kalanchoe* genus (Kolodziejczyk-Czepas and Stochmal 2017). However, as Kolodziejczyk-Czepas and Stochmal (2017) correctly pointed out, the chemistry and biological activities of bufadienolides synthesized by *Kalanchoe* plants are less known.

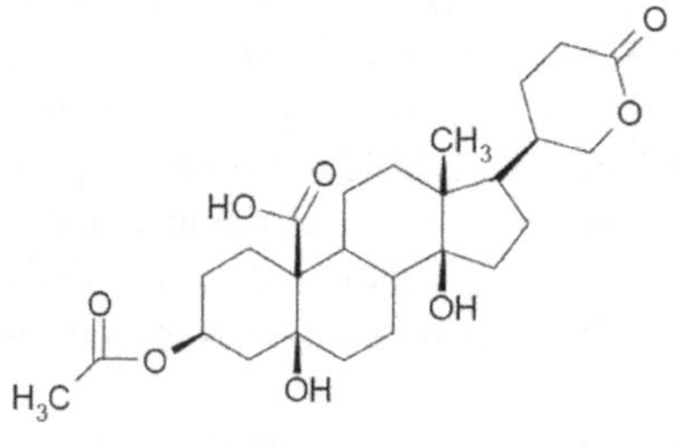

FIGURE 2.8 Chemical structure of Hellibrigenin 3-acetate isolated by Anderson et al. (1983) from *K. laciniata*, although its bioactivity has not been verified as yet.

The study by Iqbal et al. (2016) explored the antibacterial, antioxidant, and gut modulating activities of *K. laciniata* to provide a scientific rationale for its traditional uses. The methanolic extract of the plant was primarily evaluated in this investigation. In antibacterial assays the crude extract was found effective against *S. aureus* and *Bacillus subtilis*. Its antioxidant activity was also observed to be noteworthy. *K. laciniata* was one of the plants tested for its effect against skin fibroblast cell numbers by Sano et al. (2018). Positive results were observed in this instance. Aqueous-methanolic and n-hexane extracts of *K. laciniata* were evaluated for the genotoxic potential using the Ames assay and cytotoxicity was evaluated using the MTT assay in the study by Sharif et al. (2017). Both extracts were found to be cytotoxic and mutagenic. Asmah et al. (2005) investigated the proliferation effects of *K. laciniata* in hormone-dependent breast cancer (MCF-7) and colon cancer (Caco-2) cell lines. *K. laciniata* was observed to be more effective against breast cancer cells than the colon cancer cells.

Bawm et al. (2010) screened a total of 71 medicinal plant specimens from 60 plant species collected in Myanmar for anti-trypanosomal activity against trypomastigotes of *Trypanosoma evansi* and cytotoxicity against MRC-5 cells *in vitro*. *K. laciniata* was one of the plants evaluated in this study and showed potential for the treatment of *Trypanosoma evansi* infection. However, further studies were recommended in this instance, including determination and purification of active compounds that impart this beneficial effect.

Kalanchoe species contain cardiac glycosides and are toxic to animals (Smith 2004). In South Africa and Australia, where these plants are found in the wild, cattle and sheep poisonings are common (Smith 2004). Toxicosis occurs primarily in the summer months because the flowers contain a much higher concentration of glycosides than the stems, leaves, or roots (McKenzie and Dunster 1986; Reppas 1995). *Kalanchoe* species' toxicity is primarily due to a group of bufadienolide compounds, including bryotoxins, bryophyllins, and bersalgenins (Burrows and Tyrl 2001). *Kalanchoe* species' toxicosis is diagnosed based on a history of known exposure (i.e. observed ingestion, identification of chewed plants, identification of plant material in vomitus) and compatible clinical signs (Smith 2004). Although no definitive tests are available to confirm *Kalanchoe* species ingestion, assays to detect other cardiac glycosides from gastrointestinal contents have been described (McKenzie et al. 1987).

Concluding Remarks

Despite several important ethnopharmacological uses, the *K. laciniata* has not been explored extensively. Thus, it is hoped, through this chapter, that its contents will inspire future investigators for further screening of *K. laciniata* to expedite the characterization of phytoconstituents present in the plant as well as conducting *in vitro* and *in vivo* investigations along with clinical trials to verify any therapeutic effects. From a Sri Lankan perspective, it is expected that the images of the plant included herein would invoke memories in many, where the miraculous work of nature was observed by preserving leaves of *K. laciniata* in the school text books. Given the ease of propagating the plant, it is also hoped that many would initiate cultivation of

K. laciniata. From the ethnobotanical surveys, it is clear that the plant has a multitude of therapeutic properties and due attention needs to be provided to these benefits, so that they may be verified through systematic study and disseminated among potential consumers for their knowledge and awareness.

VITEX NEGUNDO LINN.

Vitex negundo Linn. is a woody and aromatic shrub. It is called 'Nika' in the vernacular language of Sinhala in Sri Lanka. It is called 'Five-Leaved Chaste Tree' or 'Monk's Pepper' in English and 'Nirgundi' in Hindi (Tandon 2005). The plant commonly bears characteristic tri- or penta-foliate leaves on quadrangular branches, which give rise to bluish-purple colored flowers in branched tomentose cymes (Vishwanathan and Basavaraju 2010). The leaf, stem, and branch structure of a young plant is shown in Figure 2.9. It thrives in humid places or along water courses in wastelands and mixed open forests and is commonly found in Afghanistan, India, Pakistan, Sri Lanka, Thailand, Malaysia, eastern Africa, and Madagascar. *V. negundo* is grown commercially as a crop in parts of Asia, Europe, North America, and the West Indies for food as well as for timber. In the US, it grows in hardiness zones and its purple flowers bloom most of the summer where they are constantly visited by bees and butterflies (Tandon 2005).

Although almost all parts of the *V. negundo* plant are used for medicinal purposes, the extracts from the leaves and the roots appear to be the most important in the traditional medicinal system of Sri Lanka. However, in several parts of the world, the pharmacological potential of *V. negundo* has been exploited effectively in formulating several commercial products for treatment of health disorders. Many of these products are readily available, especially in the consumer market of the Indian sub-continent, which contain leaf and root extracts of *V. negundo*. Most of these products are made in the form of gels for arthritis, antiseptic creams, and capsules to relieve joint pain.

FIGURE 2.9 (a) Leaf and stem structure and (b) branch structure of *V. negundo*.

A variant of *V. negundo*, is a black shrub-like tree with pointed leaves and 3–5 leaflets. This variety has small, lilac or violet flowers. The authentic black variant (referred to locally as 'Kalunika', kalu = black) is rare, and this sporadic occurrence has been incorporated into a popular metaphor of Sri Lanka, which translates as 'looking for the Kalunika', meaning searching for something that is virtually non-existent. 'Kalunika' had been from time immemorial known to the inhabitants of Sri Lanka as a powerful plant (invested with magical powers), and through lack of search, initiative, or desire, its location was neglected in modern times, even after the foreign invaders claimed to have located some trees. A twig of 'Kalunika' was believed to bring good health and fortune to those who possessed it. It is mentioned in some of the records of the colonial times of Sri Lanka, that ancient people had used a test to determine the authenticity of 'Kalunika' by throwing a small twig into the flowing water or into the fire. If the twig was the actual 'Kalunika', it was expected to move upstream and the fire to leave it intact; this, nevertheless, is something that belongs to the traditional beliefs of the country.

Origins, Morphology, and Growth

The taxonomic classification of *V. negundo* is shown in Table 2.7. It is an erect shrub or small tree growing from 2 to 8 m in height; the bark is reddish brown (Achari et al. 1984). Its leaves are digitate, with five lanceolate leaflets, sometimes three; each leaflet is around 4 to 10 cm in length, with the central leaflet being the largest and possessing a stalk (Chandramu et al. 2003). The leaf edges are toothed or serrated and the bottom surface is covered in hair. Flowers of the plant are numerous and are borne in panicles 10 to 20 cm in length; each flower is around 6 to 7 cm in length and white to blue in color (Vishwanathan and Basavaraju 2010). The petals are of different lengths, with the middle lower lobe being the longest; both the corolla and calyx are covered in dense hairs (Vishwanathan and Basavaraju 2010).

Due to the plant being exploited for commercial purposes, there are problems of adequate cultivation of *V. negundo* to meet these needs and requirements (Vishwanathan and Basavaraju 2010). It is possible, nevertheless, that *in vitro*

TABLE 2.7
Taxonomic Classification of *V. negundo*

Kingdom	Plantae
Sub-kingdom	Viridiplantae
Division	Tracheophyta
Class	Magnoliopsida
Order	Lamiales
Family	Lamiaceae
Genus	*Vitex* L.
Species	*Vitex negundo* L.

technology can potentially overcome common problems such as crop failure due to erratic weather conditions or mineral deficiencies in the soil. It is much simpler to manipulate and monitor the conditions essential for plant growth and development under laboratory conditions. Micropropagation also appears to be a prospective method for gaining rapid clonal multiplication of desired genotypes.

Traditional Medicinal Applications

During ancient times, the locals in Sri Lanka used to sleep on pillows stuffed with *V. negundo* leaves to dispel catarrh and headache and smoke the leaves for relief. A crushed leaf poultice was also applied to cure headaches, neck gland sores, tubercular neck swellings, and sinusitis. The decoction of leaves is considered as a tonic in Sri Lanka and is given along with pepper for catarrh as well as for fever. It was also used for skin irritation and inflammation (Dharmasiri et al. 2003). Many of the *Ola Leaf*-based records that denote the dosages of the herb to be used in these instances have not been located. Also, despite these traditional uses, the herb has lost some of the old applications in modern times in Sri Lanka, especially since commercial products that incorporate the herb have found their way to the marketplace of the country.

It is noteworthy that *V. negundo* has been mentioned in the verses of the *Charaka Samhita*, which is unarguably the most ancient and authoritative textbook of Indian Ayurveda (Vishwanathan and Basavaraju 2010). Patkar (2008) refers to the formulations described in the *Anubhoga Vaidya Bhaga* – a compendium of formulations in cosmetology – in outlining the use of *V. negundo* leaves along with those of *Azadirachta indica*, *Eclipta alba*, *Sphaeranthus indicus*, and *Carum copticum* in a notable rejuvenation treatment known as *Kayakalpa*. Khare (2004) outlines the applications of *V. negundo* seeds, commonly known as *Nisinda* in Unani medicine. The seeds are administered internally with sugarcane vinegar for removal of swellings. The herb is also mentioned in the Traditional Chinese Medicinal Pharmacopoeia, where the fruit of *V. negundo* is prescribed in the treatment of reddened, painful, and puffy eyes; headache; and arthritic joints (Liu et al. 2005). Although not a traditional medicinal application, the plant is used for protecting stored garlic against pests and as a cough remedy in the Philippines (Patkar 2008). In Malaysia, the herb is used in the traditional medicinal system for women's health, including treatments for regulating the menstrual cycle, fibrocystic breast disease, and post-partum remedies (Liu et al. 2005).

Bioactive Compounds of *V. negundo* and Their Therapeutic Properties

There are several chemical constituents that have been identified in various parts of *V. negundo*. Some of these are shown in Table 2.8. A potent antioxidant activity of the phenolic compounds present in the *V. negundo* leaves was observed by Kumar et al. (2010). Lakshmanashetty et al. (2010) also reported potent antioxidant activity in the ethanolic leaf extract of *V. negundo*. The methanolic root extract possessed potent snake venom neutralizing capacity.

TABLE 2.8
Phytochemical Constituents of Different Plant Parts of *V. negundo*

Part of the Plant	Phytoconstituents	References
Essential oil	δ-Guaiene; guaia-3,7-dienecaryophyllene epoxide; ethyl-hexadecenoate; α-selinene; germacren-4-ol; caryophyllene epoxide; (E)-nerolidol; β-selinene; α-cedrene; germacrene D; hexadecanoic acid; p-cymene and valencene	Khokra et al. (2008)
Leaves	Hydroxy-3,6,7,3′,4′-pentamethoxyflavone	Banerji et al. (1969)
	6′-p-Hydroxybenzoyl mussaenosidic acid; 2′-p-hydroxybenzoyl mussaenosidic acid	Sehgal et al. (1982), Sehgal et al. (1983)
	5,3′-Dihydroxy-7,8,4′-trimethoxyflavanone; 5,3′-dihydroxy-6,7,4′- trimethoxyflavanone	Achari et al. (1984)
	Viridiflorol; β-caryophyllene; sabinene; 4-terpineol; gamma-terpinene; caryophyllene oxide; 1-oceten-3-ol; globulol	Singh et al. (1999)
	Betulinic acid [3β-hydroxylup-20-(29)-en-28-oic acid]; ursolic acid [2β-hydroxyurs-12-en-28-oic acid]; n-hentriacontanol; β-sitosterol; p-hydroxybenzoic acid	Chandramu et al. (2003)
Root	2β,3α-Diacetoxyoleana-5,12-dien-28-oic acid; 2α,3α-dihydroxyoleana-5,12-dien-28-oic acid; 2α,3β-diacetoxy-18-hydroxyoleana-5,12-dien-28-oic acid; vitexin and isovitexin	Srinivas et al. (2001)
	Negundin-A; negundin-B; (+)-diasyringaresinol; (+)-lyoniresinol; vitrofolal-E and vitrofolal-F	Ul-Haq et al. (2004)
	Acetyl oleanolic acid; sitosterol; 3-formyl-4.5-dimethyl-8-oxo-5H-6,7-dihydronaphtho (2,3-b)furan	Vishnoi et al. (1983)
Seed	3β-Acetoxyolean-12-en-27-oic acid; 2α, 3α-dihydroxyoleana-5,12-dien-28-oic acid; 2β,3α-diacetoxyoleana-5,12-dien-28-oic acid; 2α, 3β-diacetoxy-18-hydroxyoleana-5,12-dien-28-oic acid	Chawla et al. (1992), Chawla et al. (1992)
	Vitedoin-A; vitedoin-B; a phenylnaphthalene-type lignan alkaloid; vitedoamine-A; five other lignan derivatives	Ono et al. (2004)

Source: Vishwanathan, A.S. and Basavaraju, R., *Euro. J. Biol. Sci.*, 3, 30–42, 2010.

The plant extract significantly antagonized the *Vipera russellii* and *Naja kaouthia* venom-induced lethal activity both in *in vitro* and *in vivo* studies (Ul-Haq et al. 2004). Ul-Haq et al. (2004) isolated two new lignans, namely negundins A and B, along with (+)-diasyringaresinol, (+)-lyoniresinol, vitrofolal E, and vitrofolal F. The structures of the new compounds were established through spectral studies. Negundin B (Figure 2.10a) showed potent inhibitory activity against lipoxygenase enzyme, while vitrofolal E (Figure 2.10b) showed moderate activity against butyryl-cholinesterase.

In the study by Diaz et al. (2003), bioassay-guided fractionation of the chloroform-soluble extract of the leaves of *V. negundo* led to the isolation of the known flavone

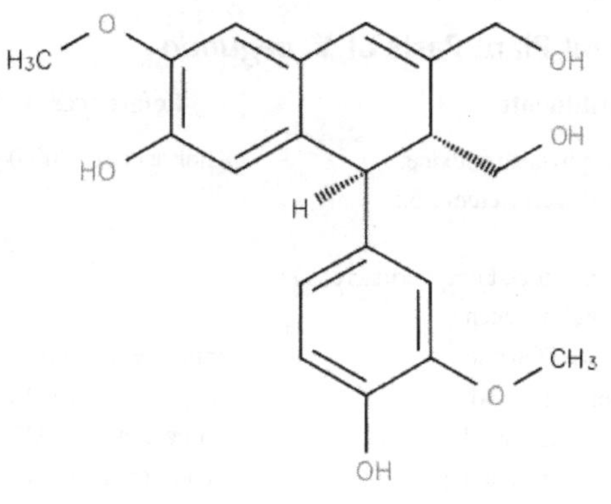

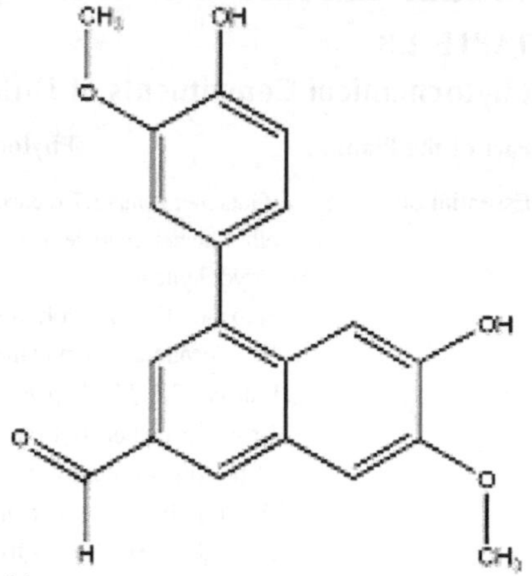

(a) (b)

FIGURE 2.10 Chemical structures of (a) negundin B and (b) vitrofolal E, which showed inhibitory activity against lipoxygenase and butyryl-cholinesterase, respectively.

vitexicarpin (Figure 2.11). This compound has shown noteworthy cytotoxic effects in human cancer cell lines. The study by Dharmasiri et al. (2003) confirmed the anti-inflammatory, analgesic, and antihistamine properties of mature fresh leaves of *V. negundo* where an aqueous extract of the leaves was orally administered to rats. Outcomes of this study revealed that the fresh leaves have anti-inflammatory and pain-suppressing activities possibly mediated via prostaglandin-synthesis inhibition, antihistamine, membrane-stabilizing, and antioxidant activities. An antihistamine activity was also demonstrated in this study, verifying the application of the herb against skin irritation in the traditional medicinal system of Sri Lanka.

Tandon et al. (2005) studied the anticonvulsant activity of *V. negundo* leaf extract in pentylenetetarazole-induced seizures in albino mice. Significant ($p < 0.05$) post-ictal depression was observed in the dose of 1000 mg/kg body weight of *V. negundo* leaf extract in comparison to the control. Tail flick tests in rats and acetic acid-induced writhing in mice were employed to study the antinociceptive activity of ethanolic leaf extract of *V. negundo* by Gupta and Tandon (2005). The leaf extract showed significant analgesic activity in a dose-dependent manner in both experimental models. It was also observed in this study that the central analgesic action does not seem to be mediated through opioid receptors.

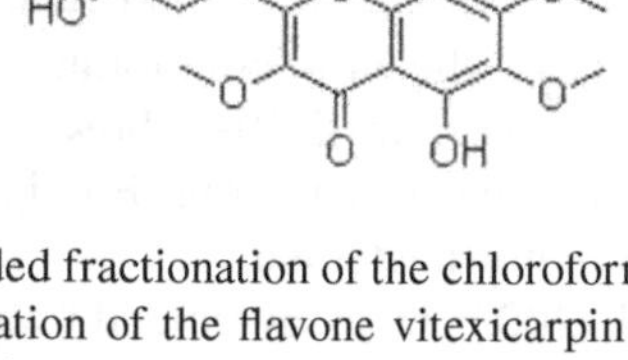

FIGURE 2.11 Bioassay-guided fractionation of the chloroform-soluble extract of the leaves of *V. negundo* led to the isolation of the flavone vitexicarpin by Diaz et al. (2003), which demonstrated cytotoxic activities.

In the study by Tandon et al. (2008), the hepatoprotective activity of *V. negundo* leaf ethanolic extract was investigated against hepatotoxicity, produced by administering a combination of three anti-tubercular drugs – isoniazid (7.5 mg/kg), rifampin (10 mg/kg), and pyrazinamide (35 mg/kg) – for 35 days by oral route in rats. Histology of the liver section of the animals treated with the *V. negundo* extract in the doses of 250 and 500 mg/kg verified the hepatoprotective activity of the herb. In the study by Tandon and Gupta (2006), carrageenan-induced hind paw edema and cotton pellet granuloma tests in albino rats were employed to study the interaction of *V. negundo* leaf extract with standard anti-inflammatory drugs in sub-effective doses to evaluate its potential role as an adjuvant therapy. The sub-effective dose of *V. negundo* potentiated anti-inflammatory activity of phenlbutazone and ibuprofen in this *in vivo* study.

The antimicrobial activity and the phytochemicals of the leaves and bark of *Vitex negundo* L. were evaluated by Panda et al. (2009) against three Gram-positive bacteria (*S. epidermidis*, *Bacillus subtilis*, and *S. aureus*) and five Gram-negative bacteria (*E. coli*, *Salmonella typhimurium*, *Pseudomonas aeruginosa*, *Vibrio cholerae*, and *Vibrio alginolyteus*). At an extract concentration of 2.5 mg/mL, 100% of inhibition was recorded against both Gram-positive and Gram-negative bacteria. *In vitro* antifungal activity of fruits of *V. negundo* was examined against five common fungal strains (*Candida albicans*, *Candida glabrata*, *Aspergillus flavus*, *Microsporum canis*, and *Fusarium solani*) in the study by Mahmud et al. (2009). A significant inhibitory activity was shown in this study against *Fusarium solani* and a moderate response against *Microsporum canis*, with no effect on *Candida albicans*.

Petroleum ether (60°C–80°C) extracts of the leaves of *V. negundo* were evaluated for larvicidal activity against larval stages of *Culex tritaeniorhynchus* in the study by Karunamoorthi et al. (2008). At a 1.5 mg/cm^2 concentration, a 6 h complete protection against mosquito bites was provided by the leaf extract, while complete protection for 8 h was found at 2.0 mg/cm^2. Overall, it was shown in this study that *V. negundo* leaf extract served as a potential larvicidal agent against Japanese encephalitis vector *C. tritaeniorhynchus* and additionally acted as a promising repellent against various adult vector mosquitoes. Anti-microfilaria effect of the methanolic extract of *V. negundo* roots was demonstrated in the *in vivo* study by Sahare et al. (2008). A complete loss of motility of the microfilariae was observed after 48 h of incubation.

Concluding Remarks

V. negundo holds much promise as a commonly available medicinal plant in Sri Lanka, and it is indeed no surprise that the plant is referred to by some of the Indian traditional medicinal practitioners as '*sarvaroganivarini*' – the remedy for all diseases. Despite the considerable amount of literature available on various aspects of the plant, there are many gaps that need to be filled through systematic study. There have been no clinical studies to date on any of the purported therapeutic properties of the plant. However, its incorporation into commercially available products does denote the safety of usage of *V. negundo.* Unlike some of the plants discussed in this book, *V. negundo* is not grown in Sri Lanka for horticultural purposes.

Thus, domestic cultivation of the plant is reduced. Yet, from the colonial records of the country and the stories and legends based on the plant, it appears that the herb is indeed invaluable and warrants scientific study to better elucidate its multitude of therapeutic effects.

OCIMUM TENUIFLORUM

O. tenuiflorum is commonly known as 'Holy Basil' in English and 'Tulasi' (sometimes spelled 'Thulasi') or 'Tulsi' in the Indian sub-continent. In Sri Lanka as well, it is known as 'Tulasi' or 'Tulsi'. It is an aromatic perennial plant in the family Lamiaceae. The plant is a native of the Indian subcontinent and is found throughout the Southeast Asian tropics as well. *O. tenuiflorum* is cultivated for religious and traditional medicinal purposes, especially to obtain its essential oil. It is widely used as an herbal tea and has a significant place within the Vaishnava tradition of Hinduism, in which devotees perform rituals involving the leaves of this plant. The leaves of *O. tenuiflorum* are an essential part in the worship of Vishnu and his avatars, including Krishna and Rama, and other male Vaishnava deities such as Hanuman, Balarama, and Garuda. Traditionally, the plant was grown in the centre of the courtyard of Hindu houses in India. It is also frequently grown next to Hanuman temples, especially in Varanasi.

Despite the religious connotations in India, in Sri Lanka the plant is mainly grown in houses for medicinal or horticultural purposes. Also, the dried leaves have been mixed with stored grains to repel insects since ancient times. A potted plant of *O. tenuiflorum* is shown in Figure 2.12. There is a particular variety of *O. tenuiflorum* used in Thai cuisine referred to as 'Thai Holy Basil' or 'Kaphrao' in Thai. One dish made with this herb is 'Phat Kaphrao' where stir-fried leaves of *O. tenuiflorum* are mixed together with meat, seafood, or rice. *O. tenuiflorum* is commonly known as 'Ruku' in Malaysia and is usually cultivated as an ornamental plant because of its small purple-yellow flower (Rabeta and Lai 2013). In Malaysia, the young leaves of *O. tenuiflorum* are used to make 'Nasi Ulam,' which is a rice-based dish (Rabeta and Lai 2013).

There appears to be several commercial products that contain extracts of *O. tenuiflorum*. For example, OciBest® is an extract of the whole plant of *O. tenuiflorum* developed by M/s Natural Remedies Pvt. Ltd., Bangalore, India (Saxena et al. 2012). This is typically consumed by and recommended for patients who are under stress. Obtaining eugenol from *O. tenuiflorum* appears to be of commercial value. In the study by Nair et al. (2012), triadimefon – a triazole compound – was used to maximize the eugenol content obtained from *O. tenuiflorum*. It was concluded in this study that triadimefon may be useful in increasing the antioxidant content in *O. tenuiflorum* and to act as an elicitor to enhance the production of its secondary metabolites.

Origins, Morphology, and Growth

The taxonomic classification of *O. tenuiflorum* is shown in Table 2.9. DNA barcodes of various biogeographical isolates of *O. tenuiflorum* from the Indian subcontinent have been carried out recently (Bast et al. 2014). In the study by Bast et al. (2014),

FIGURE 2.12 Representative image of a potted plant of *O. tenuiflorum* with its flowers.

TABLE 2.9
Taxonomic Classification of *O. tenuiflorum*

Kingdom	Plantae
Sub-kingdom	Viridiplantae
Division	Tracheophyta
Class	Magnoliopsida
Order	Lamiales
Family	Lamiaceae
Genus	*Ocimum* L.
Species	*Ocimum tenuiflorum* L.

a large-scale phylogeographical study of this species was conducted using chloroplast genome sequences, where it was found that this plant originates from north-central India. The discovery might suggest that the evolution and spread of *O. tenuiflorum* is related to the cultural migratory patterns in the Indian subcontinent. In terms of the physical characteristics of the plant, it is an erect, many-branched subshrub,

30–60 cm in height with hairy stems (Paton et al. 2004). The leaves are green or purple; they are simple, petioled, with an ovate, with a blade up to 5 cm that usually has a slightly toothed margin; they are strongly scented and have a decussate phyllotaxy (Paton et al. 2004). The purplish flowers are placed in close whorls on elongate racemes (Paton et al. 2004). The two main morphotypes cultivated in India and Nepal are green-leaved (Sri or Lakshmi Tulasi) and purple-leaved (Krishna Tulasi) (Paton et al. 2004).

The draft genome of *O. tenuiflurum* (subtype Krishna Tulsi) was presented in the report by Upadhyay et al. (2015). The paired-end and mate-pair sequence libraries were generated for the whole genome sequenced with the Illumina Hiseq 1000, resulting in an assembled genome of 374 Mb, with a genome coverage of 61%. The availability of the whole genome of *O. tenuiflorum* and the sequence analysis suggests that small amino acid changes at the functional sites of genes involved in metabolite synthesis pathways confer special medicinal properties to this herb. The expression of six important genes identified from the genome data was further validated in this study by performing q-RT-PCR in different tissues of five different species, which showed a high extent of urosolic acid-producing genes in young leaves of the Rama subtype of *O. tenuiflorum*. Out of all the herbs mentioned in this book, *O. tenuiflorum* is the only one that has been subjected to genome sequencing for elucidating its medicinal properties. Genome and transcriptome sequencing of medicinal plants serve as a robust tool for gene discovery and downstream biochemical pathway discovery of medicinally important metabolites. Recently, an abundance of transcripts for biosynthesis of terpenoids in *O. sanctum* and of phenylpropanoids in *O. basilicum* was reported during an attempt to compare transcriptomes of the two species of *Ocimum* (Rastogi et al. 2014). Nevertheless, despite having an important role in the traditional medicinal system of India and its impressive arsenal of bioactive compounds, the understanding of the biology of *O. tenuiflorum* is limited, hence the justification for the study by Upadhyay et al. (2015).

Traditional Medicinal Applications

In the traditional medicinal system of Sri Lanka, the leaves of *O. tenuiflorum* are taken as an herbal tea or dried powder, and the fresh leaf may also be consumed mixed with ghee. Comparatively, in the traditional medicinal system of India, this plant has been well documented for its therapeutic potentials and is referred to in some of the ancient traditional medicinal texts as a ‘Dashemani Shwasaharni’ and ‘Kaphaghna’ drug – both terms referring to its anti-asthmatic effects (Sirkar 1989). The plant has been used for stress and anxiety disorders in the Indian sub-continent as well (Saxena et al. 2012).

In the ethnobotanical study by Rahmatullah et al. (2009) in the Kushtia District of Bangladesh, *O. tenuiflorum* appears to be used to treat diverse illnesses such as malaria, erectile dysfunction, coughs, and colds. The juice of the crushed leaves of the plant were used to treat all these ailments. *O. tenuiflorum* was among the plants used in South Travancore, India for the treatment of skin diseases according to an ethnobotanical survey conducted by Jeeva et al. (2007). According to another ethnobotanical survey conducted by Silja, Varma, and Mohanan (2008), the Mulla Kuruma tribe of the Wayanad District of Kerala, India, used the leaf extract for treatment of skin diseases, mumps, and poisoning. Additionally, in the same tribe, steam was inhaled for coughs and colds emanating from boiled water containing leaves. In the Kanyakumari district of South India, *O. tenuiflorum* was used to treat leprosy (Kingston et al. 2009). It was found in the ethnobotanical survey by Kingston et al. (2009) that this herb was one of the exclusive materials used to treat this skin condition.

In a recent study by Manya et al. (2012), the overall prevalence and type of complementary and alternative medicines used by individuals with diabetes mellitus in western Sydney were determined. *O. tenuiflorum* appeared to be one of the herbs taken by participants in the study. It was encouraging to note that many respondents (67%) indicated they would be willing to use complementary and alternative medicines for diabetes in the future if they had positive information about the benefits from their health care providers. This would infer that individuals living with diabetes feel complementary and alternative medicines such as *O. tenuiflorum* may be a safe option and are open-minded to compliance with these therapies.

Bioactive Compounds of *O. tenuiflorum* and Their Therapeutic Properties

The chemical composition of the essential oil of the aerial flowering parts of *O. tenuiflorum* grown in northwest Karnataka, India, was investigated in the study by Joshi and Hoti (2014). Results of this study demonstrated that the oil was found to be rich in phenyl-derived compounds (83.8%). The major compound was identified as methyl eugenol (82.9%) among twenty-six compounds, comprising 98.9% of the total oil. Overall, oils extracted from the leaves and inflorescence of *O. tenuiflorum* have numerous useful properties: as expectorants, analgesics, anti-emetics, antipyretics, stress reducers, inflammation relievers, anti-asthmatic, hypoglycaemic, hepatoprotective, hypotensive, hypolipidemic and immunomodulatory (Singh, Amedkar and Verma 2010; Yamani et al. 2016). In the study by Naik et al. (2015), phytochemical screening of the leaves of *O. tenuiflorum* revealed the presence of saponins, alkaloids, flavonoids, cardiac glycosides, steroids, phenols, and tannins. In the study by Raina et al. (2013), 32 accessions of *O. tenuiflorum* germplasm were collected from different regions of northern India and were evaluated for their essential oil content and composition. The *O. tenuiflorum* oils contained phenylpropanoids, where once again, eugenol constituted the major proportion of essential oil. The range of major chemical constituents identified were eugenol (1.94%–60.20%), methyl eugenol (0.87%–82.98%), β-caryophyllene (4.13%–44.60%) and β-elemene (0.76%–32.41%).

Hexane, acetone, and methanol extracts of leaves of *O. tenuiflorum* were examined for their antimicrobial activity in this study, against selected Gram-positive and Gram-negative bacteria pathogens. The acetone extract appeared to have a wide range of antibacterial activity, whereas the methanol extract demonstrated a slightly lower antimicrobial activity than the acetone extract. Yamani et al. (2016) also verified the antimicrobial properties of essential oils distilled from Australian-grown *O. tenuiflorum* and established the quantity of the volatile components present in flower spikes, leaves, and the essential oil. The oil, at concentrations of 4.5% and

2.25%, completely inhibited the growth of *Staphylococcus aureus* and *Escherichia coli*, while the same concentrations only partly inhibited the growth of *Pseudomonas aeruginosa* in this study (Yamani et al. 2016). Out of the 54 compounds identified in the leaves of this herb, flower spikes, or essential oil, three were found responsible for this activity: camphor, eucalyptol, and eugenol. Shinde and Dhale (2011) evaluated the antifungal properties of various parts of the plant of *O. tenuiflorum*. However, it appeared that the root bark of this herb was not all that effective in inhibiting the growth of *Fusarium oxysporum* and *Rhizopus stolonifera*. Given its traditional medicinal applications, it is possible that the herb is effective in inhibiting other types of fungi.

Balaji et al. (2002) evaluated the antioxidant activities of the leaf and stem extracts of *O. tenuiflorum*. The leaf extract displayed a higher antioxidant activity than the stem, owing to the superior presence of phenolic compounds and carotenoids in the leaves. Rabeta and Lai (2013) obtained fresh leaves of *O. tenuiflorum* and subjected them to freeze-drying and vacuum-drying for processing into fermented and unfermented tea. The antioxidant activity of these samples was measured thereafter. It was observed that drying the fresh leaves of *O. tenuiflorum* significantly increased ($p < 0.05$) the antioxidant capacity, total phenolic content, total flavonoid content, and condensed tannin content.

A randomized, double-blind, placebo-controlled study was conducted by Saxena et al. (2012) to evaluate the efficacy of *O. tenuiflorum* in the symptomatic control of general stress. After six weeks of intervention, scores of symptoms such as forgetfulness, sexual problems of recent origin, frequent feeling of exhaustion, and frequent sleep problems of recent origin had decreased significantly ($p \leq 0.05$) in the herb-treated group as compared with the placebo group. Despite the availability of extensive traditional medicinal applications and traditional medicinal practitioner-based recommendations on the anti-stress activity of *O. tenuiflorum*, authenticated clinical data are still lacking. Other than the study by Saxena et al. (2012), an investigation by Bhattacharya et al. (2001) revealed the promising effects of *O. sanctum*, which belongs to the same genus, where a 500 mg capsule was administered twice a day after meals for a period of two months in the management of patients suffering from generalized anxiety disorder.

Concluding Remarks

Given the holy nature of the plant and its usage for various rites and rituals in Hinduism, it appears that the main use of the herb in Sri Lanka is primarily for medicinal purposes. In fact, the essential oil of the herb is rarely used in the traditional medicinal system of Sri Lanka, and therefore, threats of habitat loss and extinction of the plant are comparatively less in the country. As highlighted previously, clinical trials are warranted for better elucidation of the therapeutic properties of the plant, and characterization of chemical compounds that may impart other beneficial effects are waiting to be discovered as well. However, from the current research, it is very clear that the essential oil of *O. tenuiflorum* is a valuable topical antimicrobial agent for management of skin infections, and this is a possible avenue for Sri Lanka to exploit for commercial purposes.

MIMOSA PUDICA

Mimosa pudica Linn. is globally known as a weed rather than a medicinal plant (Saraiva et al. 2015). The taxonomic classification of the plant is shown in Table 2.10. Typically, when the plant is touched, the leaves fold inward and droop, and reopen within minutes (Gunawardhana et al. 2015). It is commonly known as sleeping grass, sensitive plant, humble plant, shy plant, touch-me-not, chuimui, nidikumba and lajwanti (Muhammad et al. 2016). *M. pudica* has been used in many traditional medicinal systems including Greco-Arab (Unani-Tibb), Ayurveda, and Chinese (Gilani and Atta-ur-Rahman 2005; Krishnaswamy 2008). In Sri Lanka, *M. pudica* is mainly limited to external applications such as skin conditions. However, during ancient times it was eaten in rice porridge and for select current medical issues is taken orally and described below. The plant is found growing in many gardens throughout Sri Lanka and often retained in backyards purely for medicinal applications. Being pricked by *M. pudica* thorns is common in Sri Lanka, especially among those who work on agricultural land. However, after the thorns are stripped away from the whole plant it is then applied on the skin as an antibiotic by the local population.

Origins, Morphology, and Growth

M. pudica is native to Brazil (Gunawardhana et al. 2015) and belongs to the family Mimosaceae. It is currently found in abundance in Florida, Hawaii, Virginia, Maryland, Puerto Rico, Texas, the Virgin Islands, and the South Asian subcontinent (Azmi et al. 2011; Saraswat and Pokharkar 2012). The plant has also been introduced to many other regions and is officially categorized in such areas as an invasive plant, especially in countries such as Tanzania, Southeast Asia, and many Pacific islands (Arbonnier 2004). Since it grows wild and rapidly, the spread of *M. pudica* has been officially limited in Queens Islands as well as in northern Australia (Ballabh et al. 2008). In terms of its morphology, *M. pudica* grows up to 0.5 m in height and its branches grow close to the ground and can spread around 0.3 m in length (Joseph et al. 2013). The stems are erect and slender, while the leaves are pale green,

TABLE 2.10
Taxonomic Classification of *M. pudica*

Kingdom	Plantae
Sub-kingdom	Viridiplantae
Division	Tracheophyta
Sub-division	Spermatophytina
Class	Magnoliopsida
Order	Fabales
Family	Fabaceae
Genus	*Mimosa* L.
Species	*pudica*

bipinnate, and fern-like. The leaflets exist in 15–25 pairs of usually 9–12 mm in length and 1.5 mm in width (Rocher et al. 2014). The flowers occur in globose heads and are axillary in position and lilac-pink in color (Rocher et al. 2014). The fruits are pods of 1.5–2.5 cm in length and are closely and prickly on sutures and falcate (Joseph et al. 2013). A representative image of the plant is shown in Figure 2.13.

In Sri Lanka, *M. pudica* is commonly found in moist areas, lawns, open plantations, and weedy thickets. There is no specific growing region in the country that can be identified as an area that is dense in *M. pudica* growth, since it is found in abundance in almost any place and often grows next to trees and shrubs. Because of the dense ground cover that is formed by the creeper, it prevents the reproduction of other invasive species that may be an advantage for certain types of plantations such as coconut – an important agricultural crop for Sri Lanka. In places where *M. pudica* grows wild, its growth is controlled with various chemical herbicides, especially since the plant can host parasites such as cochineal insects. *M. pudica* has thorns that may cause infections due to pricking of the skin if left untreated. Several cases of farmers being infected by *M. pudica* thorns have been reported in Sri Lanka, especially in the dry zones of the country (Gunawardhana et al. 2015).

FIGURE 2.13 *M. pudica* plant, appearing above a bush and displaying a newly blossomed lilac-colored flower. (Courtesy of Mr. Shakkya J. Ranasinghe.)

Traditional Medicinal Applications

Some of the traditional medicinal applications of *M. pudica* leaves and roots are summarized in Table 2.11. Most of these practices use either the juice, paste or powder of either the root or the leaf for the specific ailment. Other than these highlighted practices, the plant is even used to treat insomnia, irritability, pre-menstrual syndrome, menorrhagia, haemorrhoids, skin wounds, diarrhea, cough, and fever in various parts of the world (Gunawardhana et al. 2015). In Sri Lanka, the whole plant or its roots and leaves were used separately in the traditional Ayurvedic medicinal system mostly to treat skin ailments through external application only. Upon boiling the plant, the water takes on a yellowish-green hue and emanates a grassy but sweet note. The immature plant is primarily used for this purpose. When the mature plant is used, the water becomes much greener, and this is generally not recommended for topical and bathing applications on infants. Sometimes, leaves and roots are used in the treatment of piles and fistula in Sri Lanka. The leaves and roots are ground into a porridge-paste using a grindstone. In this case this paste is administered orally, generally twice a day for about 1 week.

TABLE 2.11
Traditional Medicinal Applications of *M. pudica*

Traditional Medicinal System	Part of Plant Used	Application	References
Ayurveda (India and Sri Lanka)	Root	For leprosy, dysentery, vaginal and uterine complaints, inflammations, burning sensations, asthma, leucoderma, fatigue, and blood diseases	Ediriweera and Ratnasooriya (2009), Guyton and Hall (2011)
Kandahar, India	Root	For muscular pains in the mouth, deep cuts, headaches, and stomach aches	British National Foundation (2003), Guyton and Hall (2011)
Rahba, West Bengal, India	Root	Decoction is used for gum trouble and toothaches	Nazeema and Brindha (2009), Kamboj and Kalia (2011)
Siddha System of Medicine	Leaves and root	For diabetes mellitus, glandular swellings, as eye drops to cure cataract and to treat renal colic	Ballabh et al. (2008), Tunna et al. (2014)
Unani Healthcare System	Root	To treat blood impurities and bile, bilious fevers, piles, jaundice, and leprosy	Khan (2006)
Ecuador	Leaves	Used in pillows to induce sleep in children and elderly	Van Wyk and Wink (2004), Saraiva et al. (2015)

Source: Gunawardhana, C.B. et al., *Isr. J. Plant Sci.*, 62, 234–241, 2015.

Another use of *M. pudica* is to keep away insects. Primarily, black beetles are repelled by the fragrance of the *M. pudica* flower and the crushed plant. In other parts of the Indian sub-continent, *M. pudica* is used by snake charmers against cobra bites (Mahnta and Mukherjee 2001). The stem and leaves of *M. pudica* have been traditionally applied topically against scorpion stings as well (Patwari 1992): a paste of these plant parts is applied to the skin area of the bite (Samy et al. 2008). The different traditional methods of administration or consumption of *M. pudica* between Sri Lanka and other countries of the Indian sub-continent is its use in isolation and not in combination with other medicinal herbs. In Bangladesh, for instance, *M. pudica* is ground into a pulp with onion, garlic, pepper, and saffron and administered to barren cows for the treatment of fever (Mamun et al. 2015). Piloherb – an ointment consisting of a combination of extracts of *Aloe vera*, *M. pudica*, *Azadirachta indica* (Neem), *Vitex negundo*, *Alpinia galangal*, and *Cissampelos pareira* – can be used for the treatment of haemorrhoids in India (Muhammad et al. 2016). These examples of combined preparations are not found in the Sri Lankan traditional medicinal system: the practitioners consider this herb as sufficiently powerful and potent on its own to prevent or cure the disease.

Bioactive Compounds in *M. pudica* and Their Therapeutic Effects

The traditional usage of *M. pudica* to prevent skin diseases was demonstrated in a study by Xavier et al. (2015). Antibacterial activity of *M. pudica* was demonstrated by the ability to inhibit growth of microbes that cause skin diseases. In a study by Rajan et al. (2014), the trace element content of *M. pudica* was determined in relation to the ability to cure skin diseases such as eczema, swelling, and wounds. The *M. pudica* leaf powder was shown to contain high amounts of Fe, Mn, Zn, Cu, Co, and V – elements that have been identified to be important in the maintenance of skin health.

M. pudica possesses many secondary metabolites, including carbohydrates, proteins, amino acids, tannins, phenolics, steroids, flavonoids, saponins, mucilage, and alkaloids that are of therapeutic importance, and some of the characterized compounds belonging to these categories are discussed in detail in a recent well-documented review by Muhammad et al. (2016). The major bioactive compounds present in *M. pudica* are shown in Figure 2.14. Mimosine is an important compound present in *M. pudica* and can be isolated from all parts of the plant. It is an alkaloid and is the most important compound associated with the ability of the plant to prevent the contraction of skin diseases (Restivo et al. 2005). It is also purported to possess apoptotic and, thereby, anti-cancer effects.

The roots of *M. pudica* contain endophytes that inhibit the plants from forming a symbiotic relationship (Muhammad et al. 2016). These endophytes also produce secondary metabolites such as terpenoids, steroids, alkaloids, phenols, and quinines, which defend *M. pudica* from various pathogens (Baker et al. 2012 a, b; Strobel 2003). Most traditional medicinal applications require the preparation of a decoction to extract the bioactive compounds more effectively. A decoction of *M. pudica* leaves has been reported to possess antibacterial (Balakrishnan et al. 2006) and

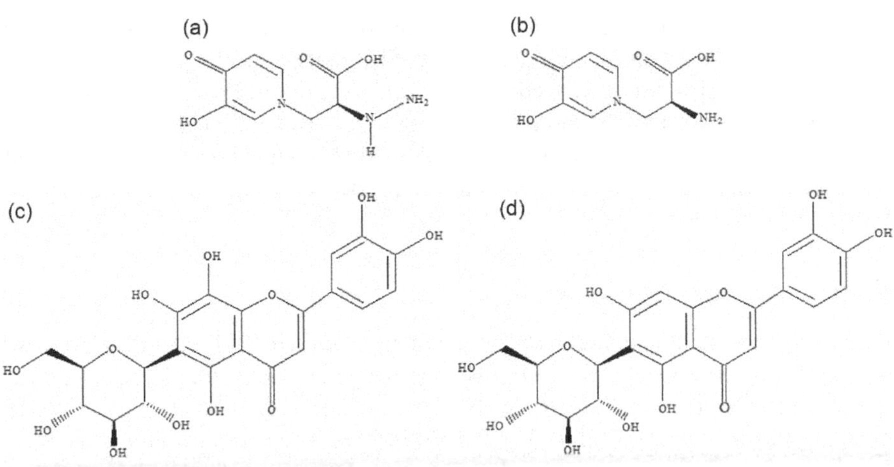

FIGURE 2.14 Bioactive compounds present in *M. pudica* (a) Mimosine amine, (b) Mimosine, (c) 3,4,7,8-tetrahydroxyl-b-D-glucopyranosyl flavone, (d) 3,4,5,7-tetrahydroxyl-b-D-glucopyranosyl flavones.

anticonvulsant activities (Bum et al. 2004). It is also used for the treatment of cough and influenza by traditional medicinal practitioners (Painkra et al. 2015).

M. pudica possesses anti-diabetic effects in *in vivo* and *in vitro* studies. Viswanathan et al. (2013) evaluated the antioxidant and antidiabetic activity of *M. pudica* in alloxan-induced diabetic rats. It was found that the aqueous extract of the whole plant inhibited the generation of superoxide, hydroxyl radical, and lipid peroxidation by the Fe^{2+}/ascorbate system as well as the Fe^{3+}/ascorbate/ADP system and nitric oxide radical. Zhang et al. (2011) evaluated the whole plant, leaves, stem, and seeds of *M. pudica* for their antioxidant activities. All parts of the plant were shown to contain significant amounts of phenolic and flavonoid compounds, with a higher antioxidant activity than the Trolox standard (Zhang et al. 2011). *M. pudica* was among the 18 plants that were investigated by Lee et al. (2015) for antioxidant and starch hydrolase inhibitory activities. The aqueous extract demonstrated radical scavenging activity along the medium range among the investigated plants. Patel and Bhutani (2014) isolated 14 compounds from the ethyl acetate extract of *M. pudica* whole plant and these compounds exhibit NO, TNF-α and IL-1β inhibitory activities in lipopolysaccharide-stimulated RAW 264.7 and J774A.1 cells. In this study, mimosine, crocetin, crocin, and jasmonic acid were identified as potent cytokine inhibitors when tested on the macrophages. Sowmya and Ananthi (2011) studied the hypolipidemic activity of the ethanolic extract of *M. pudica* on high-fat diet-induced models of hyperlipidemia in rats. This extract imparted hypolipidemic activity comparable to lovastatin – a drug that is commonly prescribed for the treatment of hyperglycaemia-induced dyslipidemia.

M. pudica possesses diuretic properties, according to a study by Baghel et al. (2013). In this study, the ethanolic extract of *M. pudica* exhibited significant diuretic activity at dosages of 100 and 200 mg/kg body weight when administered to adult albino rats. Although it is not recorded as a diuretic in the Sri Lankan traditional medicinal system, it has been used for this purpose in other countries in the Indian sub-continent. There are many pharmaceutical diuretics available such as mannitol, thiazides, furosemide, and ethacrynic acid (Gunawardhana, Ranasinghe and Waisundara 2015). However, there is a need for more effective and less-toxic diuretic agents. In view of this void, *M. pudica* could easily be singled out for more in-depth study as an herb that is able to exert diuretic properties without imparting any significant side effects.

Concluding Remarks

Regarding the potential of mass production of *M. pudica* in Sri Lanka for global use, the plant can be easily exploited for this purpose without suffering losses of extinction and genetic diversity. Essentially, the plant can be easily cultivated in any household, increasing its accessibility. Thus, there is little incentive for mass production. In terms of the bioactive compounds present in *M. pudica*, further investigations are needed, requiring large-scale isolation, particularly mimosine, and in-depth *in vivo* and toxicity studies leading to possibly a human clinical trial to elucidate the anti-carcinogenic properties. An in-depth understanding and knowledge of the distribution, intake, absorption, and metabolism of *M. pudica* plant extracts or its bioactives are of critical importance in order to fully evaluate the plant's potential benefit to human health. In light of its wide range of traditional medicinal applications in Sri Lanka, *M. pudica* is certainly a therapeutic agent that should not simply be dismissed as a common garden weed.

REFERENCES

Abbasi, A.M., M.A. Khan, M. Ahmad, M. Zafar, H. Khan, N. Muhammad and S. Sultana. 2009. Medicinal plants used for the treatment of jaundice and hepatitis based on socio-economic documentation. *African Journal of Biotechnology* 8(8):1643–1650.

Achari, B., U.S. Chowdhuri, P.K. Dutta and S.C. Pakrashi. 1984. Two isomeric flavones from *Vitex negundo. Phytochemistry* 23:703–704.

Albuquerque, U.P. 2001. The use of medicinal plants by the cultural descendants of African people in Brazil. *Acta Farmacéutica Bonaerense* 20:139–144.

Ali, H.A., O.A. Almaghrabi and M.E. Afifi. 2014. Molecular mechanisms of anti-hyperglycemic effects of *Costus speciosus* extract in streptozotocin-induced diabetic rats. *Saudi Medical Journal* 35(12):1501–1506.

Anderson, L.A.P., R.A. Schultz, J.P.J. Joubert, L. Prozesky, T.S. Kellerman, G.L. Erasmus and P.J. Krimpsiekte. 1983. Acute cardiac glycoside poisoning in sheep caused by bufadienolides from the plant *Kalanchoe lanceolata* Forsk. *Journal of Veterinary Research* 50:295–300.

Arbonnier, M. 2004. *Trees, Shrubs and Lianas of West African Dry Zones.* 2nd ed. Weikersheim, Germany: Margraf Publishers GMBH.

Ariharan, V.N., V.N. Meena Devi, M. Rajakokhila and P.N. Prasad. 2012. Antibacterial activity of *Costus speciosus* rhizome extract on some pathogenic bacteria. *International Journal of Advanced Life Sciences* 4:4–27.

Arumugam, G., M. Kumaraswamy and U.R. Sinniah. 2016. *Plectranthus amboinicus* (Lour.) Spreng: Botanical, phytochemical, pharmacological and nutritional significance. *Molecules* 21:369. doi:10.3390/molecules21040369.

Arunvanan, M., S.K. Sasi, H. Mubarak and A. Kanagarajan. 2013. An overview on anti-diabetic activity of Siddha medicinal plants. *Asian Journal of Pharmaceutical and Clinical Research* 6:46–50.

Asmah, R., M.Z.Z. Nadia, M.A. Abdah and A.B.M. Fadzelly. 2005. Effects of *Catharanthus roseus, Kalanchoe laciniatia* and *Piper longum* extracts on the proliferation of hormone-dependent breast cancer (MCF-7) and colon cancer (Caco-2) cell lines. *Malaysian Journal of Medical and Health Sciences* 1(2):105–110.

Attanayake, A.P., K.A.P.W. Jayatilaka and L.K.B. Mudduwa. 2016. Anti-diabetic potential of ivy gourd (*Coccinia grandis*, family: Cucurbitaceae) grown in Sri Lanka: A review. *Journal of Pharmacognosy and Phytochemistry* 5(6):286–289.

Attanayake, A.P., K.A.P.W. Jayatilaka, C. Pathirana and L.K.B. Mudduwa. 2015. Antihyperglycemic activity of *Coccinia grandis* (L.) Voigt in streptozotocin-induced diabetic rats. *Indian Journal of Traditional Knowledge* 14(3):376–381.

Azmi, L., M.K. Singh and A.K. Akhtar. 2011. Pharmacological and biological overview on *Mimosa pudica* Linn. *International Journal of Pharmaceutical Life Sciences* 2:1226–1234.

Baghel, A., D.S. Rathore and V. Gupta. 2013. Evaluation of diuretic activity of different extracts of *Mimosa pudica* Linn. *Pakistan Journal of Biological Sciences* 16:1223–1225.

Baker, S., S. Sahana, D. Rakshith, H.U. Kavitha, K.S. Kavitha and S. Satish. 2012a. Biodecaffeination by endophytic *Pseudomonas* sp. isolated from *Coffee arabica* L. *Journal of Pharmaceutical Research* 5(7):3654–3657.

Baker, S., P. Santhosh, D. Rakshith and S. Satish. 2012b. Screening of bacterial endophytes inhabiting *Mimosa pudica* L. *Scientific Journal of Microbiology* 1(5):136–140.

Balaji, R., G. Prakash, P. Suganya Devi and K.M. Aravinthan. 2002. Antioxidant activity of methanol extract of *Ocimum tenuiflorum* (dried leaf and stem). *International Journal of Pharmaceutical Research and Development* 3(1):20–27.

Balakrishnan, N., D. Suresh, G.S. Pandian, E. Edwin and J. Sheeja. 2006. Antidiarrhoeal potential of *Mimosa pudica* root extracts. *Indian Journal of Natural Products* 22(2):21–23.

Ballabh, B., O.P. Chaurasia, Z. Ahmed and S.B. Singh. 2008. Traditional medicinal plants of cold desert Ladakh used against kidney and urinary disorders. *Journal of Ethnopharmacology* 118:331–339.

Banerji, A., M.S. Chadha and V.G. Malshet. 1969. Isolation of 5-hydroxy-3,6,7,3′,4′-pentamethoxyflavone from *Vitex negundo. Phytochemistry* 8:511–512.

Bast, F., P. Rani and D. Meena. 2014. Chloroplast DNA phylogeography of holy basil (*Ocimum tenuiflorum*) in Indian Subcontinent. *The Scientific World Journal.* doi:10.1155/2014/847482.

Bavarva, J.H. and A.V.R.L. Narasimhacharya. 2008. Antihyperglycemic and hypolipidemic effects of *Costus speciosus* in alloxan induced diabetic rats. *Phytotherapy Research* 22:620–626.

Bawm, S., S. Tiwananthagorn, K.S. Lin, J. Hirota, T. Irie, L.L. Htun, N.N. Maw et al. 2010. Evaluation of Myanmar medicinal plant extracts for antitrypanosomal and cytotoxic activities. *Journal of Veterinary Medical Science* 72(4):525–528.

Bhandary, M.J., K.R. Chandrashekar and K.M. Kaveriappa. 1995. Medical ethnobotany of Siddis of Uttara Kannada district, Karnataka, India. *Journal of Ethnopharmacology* 47:149–158.

Bhatt, P., G.S. Joseph, P.S. Negi and M.C. Varadaraj. 2013. Chemical composition and nutraceutical potential of Indian borage (*Plectranthus amboinicus*) stem extract. *Journal of Chemistry.* doi:10.1155/2013/320329.

Bhatt, P. and P.S. Negi. 2012. Antioxidant and antibacterial activities of Indian borage (*Plectranthus amboinicus*) leaf extracts. *Food and Nutrition Sciences* 3(3):146–152.

Bhattacharya, S.K., A. Bhattacharya, K. Das, A.V. Muruganandam and K. Sairam. 2001. Further investigations on the antioxidant activity of *Ocimum sanctum* using different paradigms of oxidative stress in rats. *Journal of Natural Remedies* 1:6–16.

Birhanu, T., D. Abera and E. Ejeta. 2015. Ethnobotanical study of medicinal plants in selected Horro Gudurru Woredas, Western Ethiopia. *Journal of Biology, Agriculture and Healthcare* 5(1):83–93.

British National Foundation. 2003. *Plants: Diet and Health*. Report of a British Nutrition Foundation Task Force. G. Goldberg, editor. Oxford, UK: Blackwell Publishing.

Bum, E.N., D.L. Dawack, M. Schmutz, A. Rakotonirina, S.V. Rakotonirina, C. Portet, A. Jeker, H.R. Olpe and P. Herrling. 2004. Anticonvulsant activity of *Mimosa pudica* decoction. *Fitoterapia* 2:309–311.

Bunkrongcheap, R., N. Hutadilok-Towatana, K. Noipha, C. Wattanapiromsakul, M. Inafuku and H. Oku. 2014. Ivy gourd (*Coccinia grandis* L. Voigt) root suppresses adipocyte differentiation in 3T3-L1 cells. *Lipids in Health and Disease* 13:88. doi:10.1186/1476-511X-13-88.

Burrows, G.E. and R.J. Tyrl. 2001. *Crassulaceae – Toxic Plants of North America*. Ames, IA: Iowa State University Press, pp. 385–391.

Chandramu, C., D.M. Rao, D.G.L. Krupanandam and D.V. Reddy. 2003. Isolation, characterization and biological activity of betulinic acid and ursolic acid from *Vitex negundo* L. *Phytotherapy Research* 17:129–134.

Chang, J.M., C.M. Cheng, L.M. Hung, Y.S. Chung and R.Y. Wu. 2010. Potential use of *Plectranthus amboinicus* in the treatment of rheumatoid arthritis. *Evidence-Based Complementary and Alternative Medicine* 7(1):115–120.

Chawla, A.S., A.K. Sharma and S.S. Handa. 1992. Chemical investigation and anti-inflammatory activity of *Vitex negundo* seeds. *Journal of Natural Products* 55:163–167.

Chawla, A.S., A.K. Sharma, S.S. Handa and K.L. Dhar. 1992. A lignan from *Vitex negundo* seeds. *Phytochemistry* 31:4378–4379.

Chernetskyy, M.A. 2011. Problems in nomenclature and systematics in the subfamily Kalanchoideae (Crassulaceae) over the years. *Acta Agrobotanica* 6(4):67–74.

Chiu, Y.J., T.H. Huang, C.S. Chiu, T.C. Lu, Y.W. Chen, W.H. Peng and C.Y. Chen. 2012. Analgesic and anti-inflammatory activities of the aqueous extract from *Plectranthus amboinicus* (Lour.) Spreng. Both *In Vitro* and *In Vivo*. *Evidence-Based Complementary and Alternative Medicine*. doi:10.1155/2012/508137.

Chopra, M., A. Bhaumik, M. Prasad, K. Krishnamachary and B.R. Devi. 2018. Preliminary phytochemical screening and evaluation of *in vivo* antihyperlipidemic activity of various extracts of fruit of *Coccinia grandis* in rat model. *World Journal of Pharmacy and Pharmaceutical Sciences* 7(9):1341–1349.

Chopra, R.N., S.L. Nayar and I.C. Chopra. 1956. *Glossary of Indian Medicinal Plants*. New Delhi, India: CSIR Publications.

Das, P., D. Paik, A. Pramanik, T. De and T. Chakraborti. 2015. Antiproteolytic and leishmanicidal activity of *Coccinia grandis* (L.) Voigt leaf extract against *Leishmania donovani* promastigotes. *Indian Journal of Experimental Biology* 53:740–746.

Deb, J. and G.K. Dash. 2013. *Kalanchoe laciniata* (L) DC: A lesser known Indian medicinal plant. *International Journal of Science Inventions Today* 2:158–162.

Devi, K.N. and K. Periyanayagam. 2010. In vitro anti-inflammatory activity of *Plectranthus amboinicus* (Lour) Spreng by HRBC membrane stabilization. *International Journal of Pharmaceutical Studies and Research* 1(1):26–29.

Dharmasiri, M.G., J.R.A.C. Jayakody, G. Galhena, S.S.P. Liyanage and W.D. Ratnasooriya. 2003. Anti-inflammatory and analgesic activities of mature fresh leaves of *Vitex negundo*. *Journal of Ethnopharmacology* 87:199–206.

Diaz, F., D. Chavez, D.H. Lee, Q.W. Mi, H.B. Chai, G.T. Tan, L.B.S. Kardono et al. 2003. Cytotoxic flavone analogues of vitexicarpin, a constituent of the leaves of *Vitex negundo*. *Journal of Natural Products* 66:865–867.

Ediriweera, E.R.H.S.S. and W.D. Ratnasooriya. 2009. A review on herbs used in treatment of diabetes mellitus by Sri Lankan Ayurvedic and traditional physicians. *AYU* 30:373–391.

El-Hawary, S.S., R.H. El-Sofany, A.R. Abdel-Monem, R.S. Ashour and A.A. Sleem. 2012. Polyphenolics content and biological activity of *Plectranthus amboinicus* (Lour.) Spreng growing in Egypt (Lamiaceae). *Pharmacognosy Journal* 4(32):45–54.

Eliza, J., P. Daisy, S. Ignacimuthu and V. Duraipandiyan. 2009. Normo-glycemic and hypolipidemic effect of costunolide isolated from *Costus speciosus* (Koen ex. Retz.) Sm. in streptozotocin-induced diabetic rats. *Chemico-Biological Interactions* 179(2009):329–334.

França, F., E.L. Lago and P.D. Marsden. 1996. Plants used in the treatment of leishmanial ulcers due to *Leishmania (vlannia) braziuensis* in an endemic area of Bahia, Brazil. *Revista da Sociedade Brasileira de Medicina Tropical* 29(3):229–232.

Gaspar-Marques, C., P. Rijo, M.F. Simões, M.A. Duarte and B. Rodriguez. 2006. Abietanes from *Plectranthus grandidentatus* and *P. hereroensis* against methicillin- and vancomycin-resistant bacteria. *Phytomedicine* 13:267–271.

Gilani, A.H. and Atta-Ur Rahman. 2005. Trends in ethnopharmacology. *Journal of Ethnopharmacology* 100:43–49.

Gunawardhana, C.B., S.J. Ranasinghe and V.Y. Waisundara. 2015. *Mimosa pudica* Linn.: The garden weed with therapeutic properties. *Israel Journal of Plant Sciences* 62(4):234–241. doi:10.1080/07929978.2015.1066997.

Gupta, R.K. and V.R. Tandon. 2005. Antinociceptive activity of *Vitex negundo* Linn. leaf extract. *Indian Journal of Physiology and Pharmacology* 49(2):163–170.

Gurgel, A.P.A.D., J.G. da Silva, A.R.S. Grangeiro, H.S. Xavier, R.A.G. Oliveira, M.S.V. Pereira and I.A. de Souza. 2009. Antibacterial effects of *Plectranthus amboinicus* (Lour.) Spreng (Lamiaceae) in methicillin resistant *Staphylococcus aureus* (MRSA). *Latin American Journal of Pharmacy* 28(3):460–464.

Guyton, A. and J. Hall. 2011. *Textbook of Medical Physiology*. 12th ed. Philadelphia, PA: Elsevier.

Hasan, M.F. and B. Sikdar. Screening of antimicrobial, cytotoxic and pesticidal activities of *Coccinia grandis* (l.) Voigt. *Journal of Microbiology, Biotechnology and Food Sciences* 5(6):584–588. doi:10.15414/jmbfs.2016.5.6.584-588.

Hossain, S.A., S.N. Uddin, M.A. Salim and R. Haque. 2014. Phytochemical and pharmacological screening of *Coccinia grandis* Linn. *Journal of Scientific and Innovative Research* 3(1):65–71.

Hutchings, A., A.H. Scott, G. Lewis and A. Cunningham. 1996. *Zulu Medicinal Plants: An Inventory*. Pietermaritzburg, South Africa: University of Natal Press.

Iqbal, S.M., Q. Jamil, N. Jamil, M. Kashif, R. Mustafa and Q. Jabeen. 2016. Antioxidant, antibacterial and gut modulating activities of *Kalanchoe laciniata*. *Acta Poloniae Pharmaceutica – Drug Research* 73(5):1221–1227.

Jeeva, G.M., S. Jeeva and C. Kingston. 2007. Traditional treatment of skin diseases in South Travancore, southern peninsular India. *Indian Journal of Traditional Knowledge* 6(3):498–501.

Joseph, B., J. George and J. Mohan. 2013. Pharmacology and traditional uses of *Mimosa pudica*. *International Journal of Pharmaceutical Science and Drug Research* 5:41–44.

Joshi, R.K. and S.L. Hoti. 2014. Chemical composition of the essential oil of *Ocimum tenuiflorum* L. (Krishna Tulsi) from north west Karnataka, India. *Plant Science Today* 1(3):99–102.

Kala, C., S.S. Ali, M. Abid, U.S. Sharma and N.A. Khan. 2015. Evaluation of *in-vivo* anti-arthritic potential of methanolic extract of *Costus speciosus* rhizome. *Journal of Applied Pharmaceutical Science* 5(8):46–53.

Kaliappan, N.D. and P.K. Viswanathan. 2008. Pharmacognostical studies on the leaves of *Plectranthus amboinicus* (Lour) Spreng. *International Journal of Green Pharmacy* 2(3):182–184.

Kamboj, P. and A.N. Kalia. 2011. Evaluation of *in-vitro* (non and sitespecific) antioxidant potential of *Mimosa pudica* roots. *International Journal of Pharmaceutical Science* 3:497–501.

Karunamoorthi, K., S. Ramanujam and R. Rathinasamy. 2008. Evaluation of leaf extracts of *Vitex negundo* L. (Family: Verbenaceae) against larvae of *Culex tritaeniorhynchus* and repellent activity on adult vector mosquitoes. *Parasitology Research* 103:545–550.

Karuppuswamy, S. 2007. Medicinal plants used by Paliyan tribes of Sirumalai hills of Southern India. *Natural Product Radiance* 6:436–442.

Katulanda, P., P. Ranasinghe, R. Jayawardena, G.R. Constantine, M.R. Sherif and D.R. Mathews. 2012. The prevalence, patterns and predictors of diabetic peripheral neuropathy in a developing country. *Diabetology & Metabolic Syndrome* 4(1):21. doi:10.1186/1758-5996-4-21.

Khan, I.A. 2006. Issues to botanicals. *Life Sciences* 78:2033–2038.

Khan, M.C.P.I. 2013. *Current Trends in Coleus Aromaticus: An Important Medicinal Plant.* Bloomington, IA: Booktango.

Khare, C.P. 2004. *Encyclopedia of Indian Medicinal Plants.* Berlin, Germany: Springer.

Khokra, S., O. Prakash, S. Jain, K. Aneja and Y. Dhingra. 2008. Essential oil composition and antibacterial studies of *Vitex negundo* Linn. extracts. *Indian Journal of Pharmaceutical Sciences* 70:522–526.

Kingston, C., S. Jeeva, G.M. Jeeva, S. Kiruba, B.P. Mishra and D. Kannan. 2009. Indigenous knowledge of using medicinal plants in treating skin diseases in Kanyakumari district, Southern India. *Indian Journal of Traditional Knowledge* 8(2):196–200.

Kolodziejczyk-Czepas, J. and A. Stochmal. 2017. Bufadienolides of *Kalanchoe* species: An overview of chemical structure, biological activity and prospects for pharmacological use. *Phytochemistry Reviews* 16:1155–1171.

Kondhare, D. and H. Lade. 2017. Phytochemical profile, aldose reductase inhibitory, and antioxidant activities of Indian traditional medicinal *Coccinia grandis* (L.) fruit extract. *Biotechnology* 7:378. doi:10.1007/s13205-017-1013-1.

Krishnaswamy, K. 2008. Traditional Indian spices and their health. *Asia Pacific Journal of Clinical Nutrition* 17:265–268.

Kshirsagar, R. and A. Saklani. 2007. Ethnomedicinal plants for diabetes, Jaundice and Rheumatism in Karnataka and their comparison with Northeast India. *Advances in Ethnobotany* 95–116.

Kumar, P.P., S. Kumaravel and C. Lalitha. 2010. Screening of antioxidant activity, total phenolics and GC-MS study of *Vitex negundo*. *African Journal of Biochemistry Research* 4(7):191–195.

Laboni, F.R., T. Sultana, S. Kamal, S. Karim, S. Das, M. Harun-Or-Rashid and M. Shahrihar. 2017. Biological investigations of the ethanol extract of the aerial part (leaf) of *Coccinia grandis* L. *Journal of Pharmacognosy and Phytochemistry* 6(2):134–138.

Lakshmanashetty, R.H., V.B. Nagaraj, M.G. Hiremath and V. Kumar. 2010. *In vitro* antioxidant activity of *Vitex negundo* L. leaf extracts. *Chiang Mai Journal of Science* 37(3):489–497.

Lee, Y.H., C. Choo, M.I. Watawana, N. Jayawardena and V.Y. Waisundara. 2015. An appraisal of eighteen commonly consumed edible plants as functional food based on their antioxidant and starch hydrolase inhibitory activities. *Journal of the Science of Food and Agriculture* 95:2956–2964.

Liu, C., A. Tseng and S. Yang. 2005. *Chinese Herbal Medicine: Modern Applications of Traditional Formulas.* UK: CRC Press.

Lukhoba, C.W., M.S.J. Simmonds and A.J. Paton. 2006. *Plectranthus*: A review of ethnobotanical uses. *Journal of Ethnopharmacology* 103:1–24.

Mahanta, M. and A.K. Mukherjee. 2001. Neutralisation of lethality, myotoxicity and toxic enzymes of *Naja kaouthia* venom by *Mimosa pudica* root extracts. *Journal of Ethnopharmacology* 75:55–60.

Mahmud, M., H. Shareef, U. Farrukh, A. Kamil and G.H. Rizwani. 2009. Antifungal activities of *Vitex negundo* Linn. *Pakistan Journal of Botany* 41(4):1941–1943.

Mamun, A.A., S.I. Tumpa, I. Hossain and T. Ishika. 2015. Plant resources used for traditional ethnoveterinary phytotherapy in Jessore District, Bangladesh. *Journal of Pharmacy and Phytochemistry* 3(6):260–267.

Manan, M., L. Hussain, H. Ijaz, B. Nawaz and M. Hanif. 2015. Phytochemical screening of different extracts of *Kalanchoe laciniata* (L). *Pakistan Journal of Pharmaceutical Research* 1(2):58–61.

Manjamalai, A., T. Alexander and V.M.B. Grace. 2012. Bioactive evaluation of the essential oil of *Plectranthus amboinicus* by GC-MS analysis and its role as a drug for microbial infections and inflammation. *International Journal of Pharmacy and Pharmaceutical Sciences* 4(3):205–211.

Manya, K., B. Champion and T. Dunning. 2012. The use of complementary and alternative medicine among people living with diabetes in Sydney. *BMC Complementary and Alternative Medicine* 12:2. doi:10.1186/1472-6882-12-2.

Marles, R.J. and N.R. Farnsworth. 1995. Anti-diabetic plants and their active constituents. *Phytomedicine* 2:133–139.

McKenzie, R.A. and P.J. Dunster. 1986. Hearts and flowers: *Bryophyllum* poisoning of cattle. *Australian Veterinary Journal* 63(7):222–227.

McKenzie, R.A., F.P. Franke and P.J. Dunster. 1987. The toxicity to cattle and bufadienolide content of six Bryophyllum species. *Australian Veterinary Journal* 64(10):298–301.

Medagama, A.B., R. Bandara, R.A. Abeysekera, B. Imbulpitiya and T. Pushpakumari. 2014. Use of complementary and alternative medicines (CAMs) among type 2 diabetes patients in Sri Lanka: A cross sectional survey. *BMC Complementary and Alternative Medicine* 14(1):374. doi:10.1186/1472-6882-14-374.

Meenatchi, P., A. Purushothaman and S. Maneemegalai. 2017. Antioxidant, antiglycation and insulinotrophic properties of *Coccinia grandis* (L.) *in vitro*: Possible role in prevention of diabetic complications. *Journal of Traditional and Complementary Medicine* 7(2017):54–64.

Megersa, M., Z. Asfaw, E. Kelbessa, A. Beyene and B. Woldeab. 2013. An ethnobotanical study of medicinal plants in Wayu Tuka district, East Welega Zone of Oromia regional state, West Ethiopia. *Journal of Ethnobiology and Ethnomedicine* 9:68. doi:10.1186/1746-4269-9-68.

Modak, M., P. Dixit, J. Londhe, S. Ghaskadbi and T.P.A. Devasagayam. 2007. Indian herbs and herbal drugs used for the treatment of diabetes. *Journal of Clinical Biochemistry and Nutrition* 40:163–173.

Moosmann, B. and C. Behl. 2002. Antioxidants as treatment for neurodegenerative disorders. *Expert Opinion on Investigational Drugs* 11:1407–1435.

Muhammad, G., M.A. Hussain, I. Jantan and S.N.A. Bukhari. 2016. *Mimosa pudica* L., a high-value medicinal plant as a source of bioactives for pharmaceuticals. *Comprehensive Reviews in Food Science and Food Safety* 15:303–315.

Munasinghe, M.A.A.K., C. Abeysena, I.S. Yaddehige, T. Vidanapathirana and P.B. Piyumal. 2011. Blood sugar lowering effect of *Coccinia grandis* (L.) J. Voigt: Path for a new drug for diabetes mellitus. *Experimental Diabetes Research*. doi:10.1155/2011/978762.

Naik, S.L., P. Shyam, K.P. Marx, S. Baskari and V.R. Devi. 2015. Antimicrobial activity and phytochemical analysis of *Ocimum tenuiflorum* leaf extract. *International Journal of PharmTech Research* 8(1):88–95.

Nair, S.V.G., M. Hettihewa and H.P. Vasantha Rupasinge. 2014. Apoptotic and inhibitory effects on cell proliferation of hepatocellular carcinoma HepG2 cells by methanol leaf extract of *Costus speciosus*. *BioMed Research International*. doi:10.1155/2014/637098.

Nair, V.D., R. Gopi, M. Mohankumar, J. Kavina and R. Panneerselvam. 2012. Effect of triadimefon: A triazole fungicide on oxidative stress defense system and eugenol content in *Ocimum tenuiflorum* L. *Acta Physiologiae Plantarum* 34:599–605.

Nazeema, T.H. and V. Brindha. 2009. Antihepatotoxic and antioxidant defense potential of *Mimosa pudica*. *International Journal of Drug Discovery* 1:1–4.

Neelesh, M., J. Sanjah and M. Sappa. 2010. Anti-diabetic potentials of medicinal plants. *Acta Poloniae Pharmaceutica Drug Research* 67:113–118.

Ng, T.B., F. Liu and Z.T. Wang. 2000. Antioxidant activity of natural products from plants. *Life Sciences* 66:709–723.

Ono, M., Y. Nishida, C. Masuoka, J. Li, M. Okawa, T. Ikeda and T. Nohara. 2004. Lignan derivatives and a norditerpene from the seeds of *Vitex negundo*. *Journal of Natural Products* 67:2073–2075.

Painkra, V.K., M.K. Jhariya and A. Raj. 2015. Assessment of knowledge of medicinal plants and their use in tribal region of Jashpur district of Chhattisgarh. *Indian Journal of Applied and Natural Sciences* 7(1):434–442.

Panda, S.K., H.N. Thatoi and S.K. Dutta. 2009. Antibacterial activity and phytochemical screening of leaf and bark extracts of *Vitex negundo* L. from similipal biosphere reserve, Orissa. *Journal of Medicinal Plants Research* 3(4):294–300.

Patel, N.K. and K.K. Bhutani. 2014. Suppressive effects of *Mimosa pudica* constituents on the production of LPS-induced proinflammatory mediators. *EXCLI Journal* 13:1011–1021.

Patel, R., N.K. Mahobia, R. Gendle, B. Kaushik and S.K. Singh. 2010. Diuretic activity of *Plectranthus amboinicus* (Lour) Spreng in male albino rats. *Pharmacognosy Research* 2(2):86–88.

Patel, R.D., N.K. Mahobia, M.P. Singh, A. Singh, N.W. Sheikh, G. Alam and S.K. Singh. 2010. Antioxidant potential of leaves of *Plectranthus amboinicus* (Lour) Spreng. *Der Pharmacia Lettre* 2(4):240–245.

Patkar, K. 2008. Herbal cosmetics in ancient India. *Indian Journal of Plastic Surgery* 41:134–137.

Paton, A.J., D. Springate, S. Suddee, D. Otieno, R.J. Grayer, M.M. Harley, F. Willis, M.S.J. Simmonds, M.P. Powell and V. Savolainen. 2004. Phylogeny and evolution of basils and allies (Ocimeae, Labiatae) based on three plastid DNA regions. *Molecular Phylogenetics and Evolution* 31(1):277–299.

Patwari, B. 1992. *A Glossary of Medicinal Plants of Assam and Meghalaya*. 1st ed. Guwahati, India: M. N. Printers.

Pawar, V.A. and P.R. Pawar. 2012. *Costus speciosus*: An important medicinal plant. *International Journal of Scientific Research* 3:28–33.

Perera, H.K.I., W.K.V.K. Premadasa and J. Poonguran. 2006. α-glucosidase and glycation inhibitory effects of *Costus speciosus* leaves. *BMC Complementary and Alternative Medicine* 16:2. doi:10.1186/s12906-015-0982-z.

Rabe, T. and J. Van Staden. 1998. Screening of *Plectranthus* species for antibacterial activity. *South African Journal of Botany* 64:62–65.

Rabeta, M.S. and S.Y. Lai. 2013. Effects of drying, fermented and unfermented tea of *Ocimum tenuiflorum* Linn. on the antioxidant capacity. *International Food Research Journal* 20(4):1601–1608.

Rahman, M.A., J. Sarker, S. Akter, A.A. Mamun, M.A.K. Azad, M. Mohiuddin, S. Akter and S. Sarwar. 2015. Comparative evaluation of antidiabetic activity of crude methanolic extract of leaves, fruits, roots and aerial parts of *Coccinia grandis*. *Journal of Plant Sciences* 2(6–1):19–23.

Rahmatullah, M., D. Ferdausi, M.A.H. Mollik, M.N.K. Azam, M. Taufiq-Ur-Rahman and R. Jahan. 2009. Ethnomedicinal survey of Bheramara area in Kushtia district, Bangladesh. *American-Eurasian Journal of Sustainable Agriculture* 3(3):534–541.

Raina, A.P., A. Kumar and M. Dutta. 2013. Chemical characterization of aroma compounds in essential oil isolated from 'holy basil' (*Ocimum tenuiflorum* L.) grown in India. *Genetic Resources and Crop Evolution* 60:1727–1735.

Rajan, J.P., K.B. Singh, S. Kumar and R.K. Mishra. 2014. Trace elements content in the selected medicinal plants traditionally used for curing skin diseases by the natives of Mizoram, India. *Asian Pacific Journal of Tropical Medicine* S1:S410–S414.

Rao, G.M.M., M. Vijayakumar, C.V. Rao, A.K.S. Rawat and S. Mehrotra. 2003. Hepatoprotective effect of *Coccinia indica* against CCl4 induced hepatotoxicity. *Natural Product Science* 9(1):13–17.

Rastogi, S., S. Meena, A. Bhattacharya, S. Ghosh, R.K. Shukla, N.S. Sangwan, R.K. Lal et al. 2014. De novo sequencing and comparative analysis of holy and sweet basil transcriptomes. *BMC Genomics* 15:588. doi:10.1186/1471-2164-15-588.

Reppas, G.P. 1995. *Bryophyllum pinnatum* poisoning of cattle. *Australian Veterinary Journal* 72(11):425–427.

Restivo, A., L. Brard, C.O. Granai and N. Swamy. 2005. Antiproliferative effect of mimosine in ovarian cancer. *Journal of Clinical Oncology* 23:3200.

Revathy, J., S.S. Abdullah, and P.S. Kumar. 2014. Antidiabetic effect of *Costus speciosus* rhizome extract in alloxan induced albino rats. *Journal of Chemistry and Biochemistry* 2(1):13–22.

Rice, L.J., G.J. Brits, C.J. Potgieter and J. Van Standen. 2011. *Plectranthus*: A plant for the future? *South African Journal of Botany* 77:947–959.

Rocher, F., F. Dédaldéchamp, S. Saeedi, P. Fleurat-Lessard, J.F. Chollet and G. Roblin. 2014. Modifications of the chemical structure of phenolics differentially affect physiological activities in pulvinar cells of *Mimosa pudica* L. I. Multimode effect on early membrane events. *Plant Physiology and Biochemistry* 84:240–250.

Roshan, P., M. Naveen, P.S. Manjul, A. Gulzar, S. Anita and S. Sudarshan. 2010. *Plectranthus amboinicus* (Lour) Spreng: An overview. *Pharmaceutical Research* 4:1–15.

Sahare, K.N., V. Anandhraman, V.G. Meshram, S.U. Meshram, M.V.R. Reddy, P.M. Tumane and K. Goswami. 2008. Anti-microfilarial activity of methanolic extract of *Vitex negundo* and *Aegle marmelos* and their phytochemical analysis. *Indian Journal of Experimental Biology* 46:128–131.

Sajeev, K.K. and N. Sasidharan. 1997. Ethnobotanical observations on the tribals of Chinnar Wildlife Sanctuary. *Ancient Science Life* 16(4):284–292.

Samarakoon, K.W., H.H. Chaminda Lakmal, S.Y. Kim and Y.J. Jeon. 2013. Electron spin resonance spectroscopic measurement of antioxidant activity of organic solvent extracts derived from the methanolic extracts of Sri Lankan thebu leaves (*Costus speciosus*). *Journal of the National Science Foundation of Sri Lanka* 42(3):195–202.

Samy, R.P., M.M. Thwin, P. Gopalakrishnakone and S. Ignacimuthu. 2008. Ethnobotanical survey of folk plants for the treatment of snakebites in southern part of Tamilnadu, India. *Journal of Ethnopharmacology* 115(2):302–312.

Sano, K., T. Someya, K. Hara, Y. Sagane, T. Watanabe and R.G.S. Wijesekera. 2018. Effect of traditional plants in Sri Lanka on skin fibroblast cell number. *Data in Brief* 19:611–615.

Saraiva, M.E., A.V.R. de Alencar-Ulisses, D.A. Ribero, L.G.S. de Olivera, D.G. de Macedo, F.F.S. de Sousa, I.R.A. de Menezes, E.V.S.B. Sampaio and M.M.A. Souza. 2015. Plant species as a therapeutic resource in areas of the savanna in the state of Pernambuco, Northeast Brazil. *Journal of Ethnopharmacology* 171:141–153.

Saraswat, R. and R. Pokharkar. 2012. GC-MS studies of *Mimosa pudica*. *International Journal of PharmTech Research* 4:93–98.

Sarin, Y.K., K.L. Bedi and C.K. Atal. 1974. *Costus speciosus* rhizome as a source of Diosgenin. *Current Science* 43(18):569–570.

Saxena, R.C., R. Singh, P. Kumar, M.P.S. Negi, V.S. Saxena, P. Geetharani, J.J. Allan and K. Venkateshwarlu. 2012. Efficacy of an extract of *Ocimum tenuiflorum* (OciBest) in the management of general stress: A double-blind, placebo-controlled study. *Evidence-Based Complementary and Alternative Medicine.* doi:10.1155/2012/894509.

Sehgal, C.K., S.C. Taneja, K.L. Dhar and C.K. Atal. 1982. 2′-p-hydroxybenzoyl mussaenosidic acid, a new iridoid glucoside from *Vitex negundo. Phytochemistry* 21:363–366.

Sehgal, C.K., S.C. Taneja, K.L. Dhar and C.K. Atal. 1983. 6′-p-hydroxybenzoyl mussaenosidic acid, an iridoid glucoside from *Vitex negundo. Phytochemistry* 22:1036–1038.

Selim, S. and S. Al Jauoni. 2016. Anti-inflammatory, antioxidant and antiangiogenic activities of diosgenin isolated from traditional medicinal plant, *Costus speciosus* (Koen ex. Retz.) Sm. *Natural Product Research* 30(16):1830–1833.

Senthilkumar, A. and V. Venkatesalu. 2010. Chemical composition and larvicidal activity of the essential oil of *Plectranthus amboinicus* (Lour.) Spreng against *Anopheles stephensi*: A malarial vector mosquito. *Parasitology Research* 107:1275–1278.

Sharif, A., M.F. Akhtar, B. Akhtar, A. Saleem, M. Manan, M. Shabbir, M. Ashraf, S. Peerzada, S. Ahmed and M. Raza. 2017. Genotoxic and cytotoxic potential of whole plant extracts of *Kalanchoe laciniata* by Ames and MTT assay. *EXCLI Journal* 16:593–601.

Shenoy, S., H. Kumar, V.N. Thashma, K. Prabhu, P. Pai, I. Warrier, Somayaji, V.N. Madhav, K.L. Bairy and A. Kishore. 2012. Hepatoprotective activity of *Plectranthus amboinicus* against paracetamol-induced hepatotoxicity in rats. *International Journal of Pharmacology and Clinical Sciences* 1(2):32–38.

Shibib, B.A., L.A. Khan and R. Rahman. 1993. Hypoglycemic activity of *Coccinia indica* and *Momordica charantia* in diabetic rats: Depression of the hepatic gluconeogenic enzymes glucose-6-phosphatase and fructose-1, 6-bisphosphatase and elevation of both liver and red-cell shunt enzyme glucose-6-phosphate dehydrogenase. *Biochemical Journal* 292:267–270.

Shinde, V. and D.A. Dhale. 2011. Antifungal properties of extracts of *Ocimum tenuiflorum* and *Datura stramonium* against some vegetable pathogenic fungi. *Journal of Phytology* 3(12):41–44.

Silja, V.P., K.S. Varma and K.V. Mohanan. 2008. Ethnomedicinal plant knowledge of the *Mulla kuruma* tribe of Wayanad district, Kerala. *Indian Journal of Traditional Knowledge* 7(4):604–612.

Singh, V., S. Amdekar and O. Verma. 2010. *Ocimum sanctum* (tulsi): Bio-pharmacological Activities. *Webmed Central Pharmacology.* doi:10.9754/journal.wmc.2010.001046.

Singh, V., R. Dayal and J. Bartley. 1999. Volatile constituents of *Vitex negundo* leaves. *Planta Medica* 65:580–582.

Sirkar, N.N. 1989. Pharmacological basis of Ayurvedic therapeutics. In: C.K. Atal and B.M. Kapoor (Eds.), *Cultivation and Utilization of Medicinal Plants.* New Delhi, India: PID, CSIR.

Sivaraj, A., B. Preethi Jenifa, M. Kavitha, P. Inbasekar, B. Senthilkumar, N.P. Tamilselvan. 2011. Antibacterial activity of *Coccinia grandis* leaf extract in selective bacterial strains. *Journal of Applied Pharmaceutical Science* 1(7):120–123.

Smith, G. 2004. *Kalanchoe* species poisoning in pets. *Veterinary Medicine* 934–936. https://www.aspcapro.org/sites/default/files/v-vetm1104_933–936.pdf (last accessed on 11 November 2018).

Sowmya, A. and T. Ananthi. 2011. Hypolipidemic activity of *Mimosa pudica* Linn. on butter-induced hyperlipidemia in rats. *Asian Journal of Research and Pharmacy Science* 1:123–126.

Srinivas, K.K., S.S. Rao, M.E.B. Rao and M.B.V. Raju. 2001. Chemical constituents of the roots of *Vitex negundo. Indian Journal of Pharmaceutical Sciences* 63:422–424.

Staples, G.W. and M.S. Kristiansen. 1999. *Ethnic Culinary Herbs: A Guide to Identification and Cultivation in Hawaii*. Honolulu, HI: University of Hawaii Press.

Strobel, G.A. 2003. Endophytes as sources of bioactive products. *Microbes Infections* 5:535–544.

Susanti, H., S. Wahyuono, I.P. Sari and R.A. Susidarti. 2018. Antihypercholesterol activity of *Costus speciosus* water extract. *Thai Journal of Pharmaceutical Sciences* 42(2):66–68.

Tamilselvan, N., T. Thirumalai, E.K. Elumalai, R. Balaji and E. David. 2011. Pharmacognosy of *Coccinia grandis*: A review. *Asian Pacific Journal of Tropical Biomedicine* 5:299–302.

Tandon, V.R. 2005. Medicinal uses and biological activities of *Vitex negundo. Natural Product Radiance* 4(3):162–165.

Tandon, V.R. and R.K. Gupta. 2005. An experimental evaluation of anticonvulsant activity of *Vitex negundo. Indian Journal of Physiology and Pharmacology* 49(2):199–205.

Tandon, V.R. and R.K. Gupta. 2006. *Vitex negundo* Linn (VN) leaf extract as an adjuvant therapy to standard anti-inflammatory drugs. *Indian Journal of Medical Research* 124:447–450.

Tandon, V.R., V. Khajuria, B. Kapoor, D. Kour and S. Gupta. 2008. Hepatoprotective activity of *Vitex negundo* leaf extract against anti-tubercular drugs induced hepatotoxicity. *Fitoterapia* 79:533–538.

Trease, G.E. and W.C. Evans. 1978. *Pharmacology*. 1st ed. London, UK: Bailliere Tindall.

Tunna, T.S., Q.U. Ahmed, A.B.M. Helal Uddin and M.Z.I. Sarker. 2014. Weeds as alternative useful medicinal source: *Mimosa pudica* Linn. on diabetes mellitus and its complications. *Advanced Material Research* 995:49–59.

Ul-Haq, A., A. Malik, I. Anis, S.B. Khan, E. Ahmed, Z. Ahmed, S.A. Nawaz and M.I. Choudhary. 2004. Enzyme inhibiting lignans from *Vitex negundo. Chemical and Pharmaceutical Bulletin* 52:1269–1272.

Uma, M., S. Jothinayaki, S. Kumaravel and P. Kalaiselvi. 2014. Determination of bioactive components of *Plectranthus amboinicus* Lour by GC–MS analysis. *New York Science Journal* 4(8):66–69.

Upadhyay, A.K., A.R. Chacko, A. Gandhimathi, P. Ghosh, K. Harini, A.P. Joseph, A.G. Joshi et al. 2015. Genome sequencing of herb Tulsi (*Ocimum tenuiflorum*) unravels key genes behind its strong medicinal properties. *BMC Plant Biology* 15:212. doi:10.1186/s12870-015-0562-x.

Vaishnav, M.M., P. Jain, S.R. Jogi and K.R. Gupta. 2001. Coccinioside-K, triterpenoid saponin from *Coccinia indica. Oriental Journal of Chemistry* 17(3):465–468.

Van Jaarsveld, E. 2006. *South African Plectranthus and the Art of Turning Shade to Glade*. Simon's Town, South Africa: Fernwood Press.

Van Wyk, B.E. and M. Wink. 2004. *Medicinal Plants of the World*. 1st ed. Pretoria, South Africa: Briza.

Vijayakumar, S., G. Vinoj, B. Malaikozhundan and S.S.B. Vaseeharan. 2015. *Plectranthus amboinicus* leaf extract mediated synthesis of zinc oxide nanoparticles and its control of methicillin resistant *Staphylococcus aureus* biofilm and blood sucking mosquito larvae. *Spectrochimica Acta Part A: Molecular and Biomolecular Spectroscopy* 137:886–891.

Vijayalakshmi, M.A. and N.C. Sarada. 2008. Screening of *Costus speciosus* extracts for antioxidant activity. *Fitoterapia* 79(3):197–198.

Vishnoi, S.P., A. Shoeb, R.S. Kapil and S.P. Popli. 1983. A furanoeremophilane from *Vitex negundo. Phytochemistry* 22:597–598.

Vishwanathan, A.S. and R. Basavaraju. 2010. A review on *Vitex negundo* L. – A medicinally important plant. *European Journal of Biological Sciences* 3(1):30–42.

Viswanathan, R., V. Sekar, V. Velpandian, K.S. Sivasaravanan and S. Ayyasamy. 2013. Antidiabetic activity of thottal vadi choornam (*Mimosa pudica*) in alloxan induced diabetic rats. *International Journal of Natural Product Science* 3:13–20.

Vishwanathaswamy, A.H.M., B.C. Koti, A. Gore, A.H.M. Thippeswamy and R.V. Kulkarni. 2011. Anti-hyperglycemic and anti-hyperlipidemic activity of *Plectranthus amboinicus* on normal and alloxan-induced diabetic rats. *Indian Journal of Pharmaceutical Science* 73(2):139–145.

Waisundara, V.Y. and M.I. Watawana. 2014. Evaluation of the antioxidant activity and additive effects of traditional medicinal herbs from Sri Lanka. *Australian Journal of Herbal Medicine* 26(1):22–28.

Waisundara, V.Y., M.I. Watawana and N. Jayawardena. 2015. *Costus speciosus* and *Coccinia grandis*: Traditional medicinal remedies for diabetes. *South African Journal of Botany* 98(2015):1–5.

Wasantwisut, E. and T. Viriyapanich. 2003. Ivy gourd (*Coccinia grandis* Voigt, *Coccinia cordifolia, Coccinia indica*) in human nutrition and traditional applications. *World Reviews in Nutrition and Diet* 91:60–66.

Xavier, T.F., M. Kannan and A. Auxilia. 2015. Traditional medicinal plants used in the treatment of different skin diseases. *International Journal of Current Microbiology and Applied Science* 4:1043–1053.

Yamani, H.A., E.C. Pang, N. Mantri and M.A. Deighton. 2016. Antimicrobial activity of tulsi (*Ocimum tenuiflorum*) essential oil and their major constituents against three species of bacteria. *Frontiers in Microbiology*. doi:10.3389/fmicb.2016.00681.

Zhang, J., K. Yuan, W.L. Zhou, J. Zhou and P. Yang. 2011. Studies on the active components and antioxidant activities of the extracts of *Mimosa pudica* Linn. from southern China. *Pharmacognosy Magazine* 7:35–39.

3 Important History of Incorporation of Medicinal Plants into Porridge in Sri Lanka

HERBAL PORRIDGES IN MODERN TIMES

Importantly, the most common form of consuming herbs in Sri Lanka nowadays is to convert or add them into porridge. Consuming herbal porridges or 'kola kanda' (as known in Sinhala) for breakfast has become a trend in modern Sri Lanka to preserve good health as well as to obtain necessary phytonutrients. This trend is growing with increasing amounts of scientific evidence provided on Sri Lankan herbs. The younger generation find the habit of consuming herbal porridge more palatable since some of the herbs help to prevent the occurrence of diseases or even cure them according to Sri Lankan Ayurvedic medicine. When rice gruel is incorporated with minced leaves of herbs and consumed as a broth, the bitterness and grassy note present in many of the herbs is masked, and the resulting product is more appetizing. Additionally, drinking one glass of herbal porridge is considered an all-in-one breakfast among Sri Lankans, and given the presence of fibres contributed by the broth, it offers a feeling of satiety. It was shown in a study by Senadheera et al. (2015) that porridges are considered more palatable and fulfilling compared to a simple water extract of herbs or tablets and could be ingested as a meal. This is very important for the urban population of Sri Lanka, who leads high-powered corporate-level careers and do not appear to have the time to consume the traditional rice and curry breakfast.

Consuming herbal porridges as a habit follows significant history, almost as old as the Ayurvedic system of the country. Both then and now, herbal porridges were recognized as a practice that imparts health and wellness and the reasons why it has become popular at present are the scientific findings in support of the herbs that are used to prepare the porridges. According to ancient texts such as the Deepavamsa and Mahavamsa, which document the history of the country, the habit of consuming herbal porridge appears to have originated through Buddhist culture. It was especially consumed in the mornings by Buddhist priests as means of sustenance. As part of the Buddhist monastic code, the priests only take two meals a day (breakfast and lunch) and following a period without consuming solid food after lunch of the previous day, herbal porridge provides restoration and rejuvenation to start off their daily activities. This practice is followed by lay people as well. Whether the habit of consuming herbal porridge was an ancient practice among locals in other countries of the Indian sub-continent is not evident due to the unavailability of documents. Thus, until substantial proof is found, it could be mentioned that consuming herbal porridge is a habit that is unique to Sri Lankans both then and now.

HISTORY OF THE HABIT OF CONSUMING PORRIDGE IN SRI LANKA

Ancient stone inscriptions in Sri Lanka mention the phrase 'hambu bath' or 'yāguwa', which refers to rice-based gruel. These texts were written in Brahmi script and several of them are found mostly in areas around temples and lakes. An example of such a script is found in the Lankathilake temple in Pilimathalawa, Sri Lanka and shown in Figure 3.1. The construction workers who built the temples and lakes during ancient times had consumed the rice gruel for nourishment during the creation of these monuments and structures, and they made a point to mention the food, which they consumed during the construction process when carving on the stone inscriptions. The Maha Viharaya, which was one of the largest Buddhist monasteries for Theravada Buddhism in Anuradhapura, Sri Lanka during 247–207 BCE, appears to have had 'boats' of herbal rice gruel constructed out of stone within the monastery complex (Collins 2000). It is mentioned in some of the stone inscriptions surrounding the location of this monastery that these vessels, which held herbal rice gruel, were constantly filled by lay patrons of the monastery and some of these receptacles are even seen today (albeit in ruins). Sometimes the gruel may be consumed together with bees' honey, ghee, or jaggery (a traditional non-centrifugal cane sugar made of either sugar cane or palm sap).

FIGURE 3.1 Stone inscription found in the Lankathilake temple located in Pilimathalawa, Sri Lanka. Many such texts are found adjacent to ancient temples documenting the construction process of the buildings.

A prominent Buddhist priest of Sri Lanka was Thotagamuwe Sri Rahula Thero (1408–1491). Not only was he an eminent scholar (who was well versed in six languages and referred to as 'Shad Bhasha Parameshwara'), he was a distinguished author and a very proficient Ayurvedic physician (Himbutana 2006). Legend has it that because of an Ayurvedic drug made entirely out of local herbs named 'Siddāloka Rasaya' that was eaten by the Thero, his body did not decompose, and thus, due to this saintly nature, a cremation was not carried out on the cadaver. Instead, it was kept in his residence (Vijayaba Pirvena) after his death for some time, until the arrival of Portuguese invaders in 1505. The residents, who feared a possible abduction of the Thero's body by the Portuguese, took it to the Indurigiri cave for safety at Ambana near Elpitiya in the Galle district (Southern Province of Sri Lanka), which was surrounded by a dense forest cover. Sri Rahula Thero was a huge advocate of the consumption of herbal rice gruel and it was mentioned that he praised those who incorporated this food item to their daily meals, especially owing to its humane (vegetarian) nature, since it does not incur the intentional killing of another being to obtain its ingredients. A belief, which would have originated due to Sri Rahula Thero's campaign was that the habit of drinking herbal porridge provided long life, beauty, health, wellness, power, insight, and intellect.

TYPES OF RICE IN HERBAL PORRIDGES

It would be remiss to avoid describing – at least in brief – the types of rice used to prepare the herbal porridges. Several types of rice are used in Sri Lanka, where the red and white raw rice are the more common varieties (Figure 3.2). These varieties are commercially available. The rice may be cooked in the usual manner by boiling with excess amounts of water for the broth to be formed, or the grains may be browned in a pan without the addition of oil and added to excess amounts of boiling water. Either scraped coconut or coconut milk with a pinch of salt may be added to the broth in either instances to add flavor as well as impart palatability. A study by Senadheera and Ekanayake (2013) that was conducted on porridges prepared

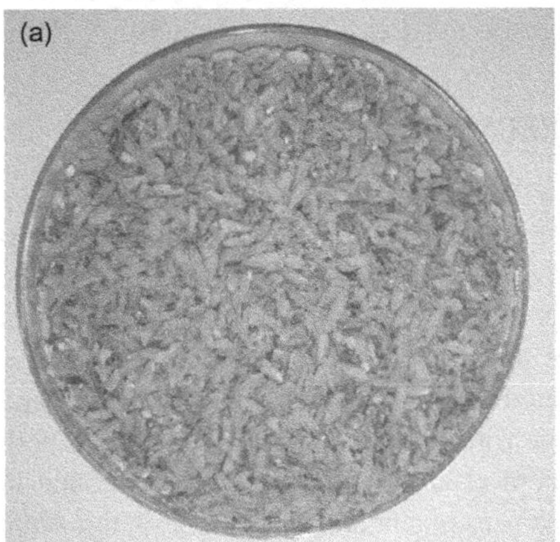

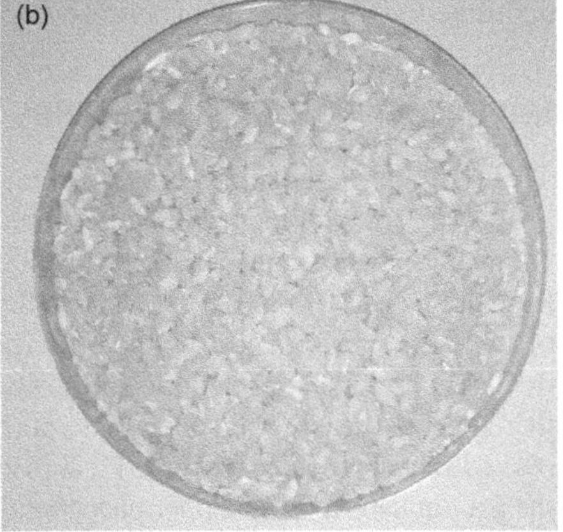

FIGURE 3.2 (a) Red raw rice and (b) white raw rice, which are typically used in Sri Lankan households nowadays for the preparation of herbal porridge.

according to the Sri Lankan style (rice:leaves:scraped coconut kernel in 25:15:10 (w/w/w) ratio), provided evidence that most of them elicit a low glycaemic index in normal subjects. This low glycaemic index is mainly due to the high-water content and not due to the hypoglycaemic effects of the added herbs (Senadheera and Ekanayake 2012). Certain types of rice porridge may have other grains added to them (such as mung beans or cow pea), or even other sweetening agents such as honey or jaggery.

PREPARATION OF HERBAL PORRIDGE

There are two recipes for the preparation of herbal porridge as given in the comprehensive Sinhala text on herbal porridges by Dr. Piyasiri Amilasith Yapa (2015). The first recipe mentions the following ingredients and their amounts:

Rice grains – One fistful per person
Herb of selection – One fistful
Grated coconut – One fistful
Salt – As required for taste
Garlic – 6 cloves
Raw ginger – 1–2 slices

The rice is ground in a grindstone until broken into somewhat small pieces. The selected herb is also ground in a grindstone in a similar manner, until the juice is extracted. Milk obtained from the grated coconut is added to a pot. Approximately 3 cups of water are added to the broken rice grains in a pot and boiled, while the coconut milk and garlic are added. The mixture is continuously stirred until the rice becomes soft. The herb extract is stirred in last together with salt, and the heat is removed so that the nutrients of the herb extract do not get destroyed. A more urban method of preparation is referenced in the book by Yapa (2015). The same ingredients are used, with grinding carried out with a blender carefully to avoid over-grinding the rice into powder. The porridge is best made with some solid particles and fibre to stimulate and encourage the digestion process in the stomach.

IMPORTANT MEDICINAL HERBS PREPARED AS A PORRIDGE

There are many medicinal herbs in Sri Lanka that are prepared in porridge form. They possess a variety of therapeutic effects from anti-diabetic to anti-cancer to anti-thrombogenic. Some examples of herbs that are traditionally used in Sri Lanka in porridge form are listed in Table 3.1. The scientifically elucidated and select therapeutic effects of the herbs are shown in Table 3.2. There are several other herbs not included in the list that are being scientifically investigated at present for their compatibility and therapeutic efficacy when added into porridge form. *Scoparia dulcis* is one such herb (Senadheera and Ekanayake 2013; Senadheera et al. 2015). Although mentioned in the Sri Lankan Ayurvedic pharmacopoeia as a very potent anti-diabetic herb, this plant was not traditionally consumed in porridge form. Thus, the list in Table 3.1 primarily focuses on herbs listed in ancient Ayurvedic texts as being incorporated into 'yāgu' or 'kanda'. The therapeutic effects of some of these herbs in Table 3.1 are elaborated

TABLE 3.1
Herbs of Sri Lanka Prepared in Porridge Form: Scientific Name, Family, Vernacular Name, Traditional Medicinal Applications and Representative Images of Some of These Conventionally Documented Herbs Used in Past and Present Times

Scientific Name	Family	Vernacular Name (Sinhala)	Traditional Medicinal Applications	Representative Image of Plant
Abrus precatorius Linn	Liquiminose	Olinda	As a blood purifier, for glandular conjunctivitis, worm infections, rabies and tetanus	
Abutilon indicum Sweet	Malvaceae	Anodā	Used for the treatment of toothaches, intestinal worms, urinary infections, bladder stones, haematuria, dysentery and disorders in the female and male reproductory systems	

(Continued)

TABLE 3.1 (*Continued*)
Herbs of Sri Lanka Prepared in Porridge Form: Scientific Name, Family, Vernacular Name, Traditional Medicinal Applications and Representative Images of Some of These Conventionally Documented Herbs Used in Past and Present Times

Scientific Name	Family	Vernacular Name (Sinhala)	Traditional Medicinal Applications	Representative Image of Plant
Acalypha indica Linn	Euphorbiaceae	Kuppameniya	As a diuretic, expectorant, emetic and carminative agent, for gastrointestinal irritations, skin diseases, constipation, fever and swellings	
Achyranthes aspera Linn	Amaranthaceae	Rathkaral	For haemorrhoids, snake bites, joint swellings, stomach aches, oral infections and as a blood purifier	

(*Continued*)

TABLE 3.1 (*Continued*)
Herbs of Sri Lanka Prepared in Porridge Form: Scientific Name, Family, Vernacular Name, Traditional Medicinal Applications and Representative Images of Some of These Conventionally Documented Herbs Used in Past and Present Times

Scientific Name	Family	Vernacular Name (Sinhala)	Traditional Medicinal Applications	Representative Image of Plant
Aerva lanata Linn	Amaranthaceae	Pol palā	For worm infections, urinary infections, cholera, diarrhea, bladder stones and gonorrhea, as a demulcent and antidote for poisoning	
Alangium salvifolium Linn	Alangiaceae	Ruk anguna	As an astringent, pungent, purgative, emetic, antipyretic and diaphoretic, laxative agent and antidote, for hypertension and diabetes	

(*Continued*)

TABLE 3.1 (*Continued*)

Herbs of Sri Lanka Prepared in Porridge Form: Scientific Name, Family, Vernacular Name, Traditional Medicinal Applications and Representative Images of Some of These Conventionally Documented Herbs Used in Past and Present Times

Scientific Name	Family	Vernacular Name (Sinhala)	Traditional Medicinal Applications	Representative Image of Plant
Alternanthera sessilis Linn	Amaranthaceae	Mukunuwenna	For chronic headaches, night blindness, eye infections, asthma, catarrh, urinary infections, fatigue, rabies, snake bites and spider bites	
Asparagus racemosus Wild	Lilaceae	Hāthāwāriya	For haemorrhoids, urinary infections, oral infections, tuberculosis, night blindness, catarrh, diabetes and rheumatism	

(*Continued*)

TABLE 3.1 (*Continued*)
Herbs of Sri Lanka Prepared in Porridge Form: Scientific Name, Family, Vernacular Name, Traditional Medicinal Applications and Representative Images of Some of These Conventionally Documented Herbs Used in Past and Present Times

Scientific Name	Family	Vernacular Name (Sinhala)	Traditional Medicinal Applications	Representative Image of Plant
Astercantha longifolia	Acanthaceae	Neeramulliya	For urinary infections, rheumatism, tonsillitis and disorders in the sexual organs in both males and females and gonorrhea	
Atlantia ceylanica Linn	Rutaceae	Yakināran	For catarrh, asthma, rheumatism, skin diseases and irritations and respiratory disorders, to cure bee and wasp stings	

(*Continued*)

TABLE 3.1 (*Continued*)
Herbs of Sri Lanka Prepared in Porridge Form: Scientific Name, Family, Vernacular Name, Traditional Medicinal Applications and Representative Images of Some of These Conventionally Documented Herbs Used in Past and Present Times

Scientific Name	Family	Vernacular Name (Sinhala)	Traditional Medicinal Applications	Representative Image of Plant
Cardiospermum microcarpum HBK Nov	Sapindaceae	Wel penela	As an emetic, laxative, tonic, diaphoretic, sedative, analgesic, vaso-depressant and demulcent, for haemorrhoids and treatment of fractures, obesity prevention	
Cassia auriculata Linn	Leguminosae	Ranawarā	For urinary infections, eye infections, swellings, cancer and diabetes, to prevent giddiness, as an emetic, purgative and diuretic	

(*Continued*)

TABLE 3.1 (*Continued*)
Herbs of Sri Lanka Prepared in Porridge Form: Scientific Name, Family, Vernacular Name, Traditional Medicinal Applications and Representative Images of Some of These Conventionally Documented Herbs Used in Past and Present Times

Scientific Name	Family	Vernacular Name (Sinhala)	Traditional Medicinal Applications	Representative Image of Plant
Ceiba pentandra Linn	Bombacaceae	Imbul	For urinary infections, gonorrhea, cold, cough, asthma, bladder stones and constipation	
Centella asiatica Urb	Umbelliferae	Gotukola	Treatment for catarrh, asthma, eye infections, spider bites, snake bites, haemorrhoids, headaches, sore throat, cracked lips, improving memory	

(*Continued*)

TABLE 3.1 (*Continued*)

Herbs of Sri Lanka Prepared in Porridge Form: Scientific Name, Family, Vernacular Name, Traditional Medicinal Applications and Representative Images of Some of These Conventionally Documented Herbs Used in Past and Present Times

Scientific Name	Family	Vernacular Name (Sinhala)	Traditional Medicinal Applications	Representative Image of Plant
Clitoria ternatea Linn	Leguminosae	Elakatarolu	For rheumatoid arthritis, rheumatism, urinary infections, eye infections and hemicrania	
Dregea volubilis Linn	Asclepediaceae	Kiri anguna	As an expectorant, emetic and tonic, for rabies, abscess, tumors, asthma, headaches and hypertension	

(*Continued*)

TABLE 3.1 (*Continued*)

Herbs of Sri Lanka Prepared in Porridge Form: Scientific Name, Family, Vernacular Name, Traditional Medicinal Applications and Representative Images of Some of These Conventionally Documented Herbs Used in Past and Present Times

Scientific Name	Family	Vernacular Name (Sinhala)	Traditional Medicinal Applications	Representative Image of Plant
Eclipta prostrata Linn	Compositae	Keekirindiya	For cough, fever, cold, wasp and bee stings, as a blood purifier and diuretic and to maintain blood pressure and prevent miscarriages and worm infections	
Erythrina variegata Linn	Leguminosae	Erabadu	To cure bee and wasp stings, toothaches, ear aches, skin diseases, prevent asthma and worm infections	

(*Continued*)

TABLE 3.1 (*Continued*)

Herbs of Sri Lanka Prepared in Porridge Form: Scientific Name, Family, Vernacular Name, Traditional Medicinal Applications and Representative Images of Some of These Conventionally Documented Herbs Used in Past and Present Times

Scientific Name	Family	Vernacular Name (Sinhala)	Traditional Medicinal Applications	Representative Image of Plant
Evolvulus alsinoides Linn	Convolvulaceae	Vishnukrantiya	To improve memory power, for diarrhea, gastrointestinal disorders, neurological disorders, bronchitis and to stimulate hair growth	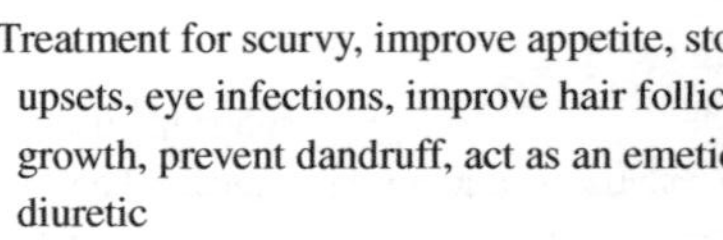
Feronia limonia Linn	Rataceae	Divul	Treatment for scurvy, improve appetite, stomach upsets, eye infections, improve hair follicle growth, prevent dandruff, act as an emetic and diuretic	

(*Continued*)

TABLE 3.1 (*Continued*)
Herbs of Sri Lanka Prepared in Porridge Form: Scientific Name, Family, Vernacular Name, Traditional Medicinal Applications and Representative Images of Some of These Conventionally Documented Herbs Used in Past and Present Times

Scientific Name	Family	Vernacular Name (Sinhala)	Traditional Medicinal Applications	Representative Image of Plant
Ficus hispida Linn	Moraceae	Kota dimbula	As a tonic, emetic and diuretic, for skin diseases and snake bites	
Ficus religiosa Linn	Moraceae	Bo	To rectify fertility issues in females, as emetic, anti-microbial agent and stool softener, for skin diseases, cholera, fever, gonorrhea, toothaches and asthma	

(*Continued*)

TABLE 3.1 (*Continued*)
Herbs of Sri Lanka Prepared in Porridge Form: Scientific Name, Family, Vernacular Name, Traditional Medicinal Applications and Representative Images of Some of These Conventionally Documented Herbs Used in Past and Present Times

Scientific Name	Family	Vernacular Name (Sinhala)	Traditional Medicinal Applications	Representative Image of Plant
Hemidesmus indicus Linn	Asclepediaceae	Iramusu	As a blood purifier, tonic, demulcent, emetic, diuretic, and as a remedy for filariasis, hiccups and snake bites	
Lasia spinosa Linn	Araceae	Kohila	For haemorrhoids and stomach aches, as a blood purifier	

(*Continued*)

TABLE 3.1 (*Continued*)

Herbs of Sri Lanka Prepared in Porridge Form: Scientific Name, Family, Vernacular Name, Traditional Medicinal Applications and Representative Images of Some of These Conventionally Documented Herbs Used in Past and Present Times

Scientific Name	Family	Vernacular Name (Sinhala)	Traditional Medicinal Applications	Representative Image of Plant
Murraya koenigi Spreng	Rutaceae	Karapinchā	For snake bites, insect bites, improving appetite, preventing liver diseases, worm infections, skin infections, diabetes and weight loss, as an emetic and for improving digestion, preventing untimely growth of grey hairs	
Osbeckia octandra DC Prodr	Melastomaceae	Heen bovitiya	For liver-related diseases, prevention of infections and skin diseases	

(*Continued*)

TABLE 3.1 (*Continued*)

Herbs of Sri Lanka Prepared in Porridge Form: Scientific Name, Family, Vernacular Name, Traditional Medicinal Applications and Representative Images of Some of These Conventionally Documented Herbs Used in Past and Present Times

Scientific Name	Family	Vernacular Name (Sinhala)	Traditional Medicinal Applications	Representative Image of Plant
Pandanus tectorius Soland	Pandanaceae	Wetaké	To prevent untimely growth of grey hair, for skin diseases, insomnia, leprosy, heart ailments and neurological disorders	
Phyllanthus debilis	Zuphoribiaceae	Sudu pitawakka	For respiratory disorders, asthma, catarrh, joint swellings and headaches	

(*Continued*)

TABLE 3.1 (*Continued*)

Herbs of Sri Lanka Prepared in Porridge Form: Scientific Name, Family, Vernacular Name, Traditional Medicinal Applications and Representative Images of Some of These Conventionally Documented Herbs Used in Past and Present Times

Scientific Name	Family	Vernacular Name (Sinhala)	Traditional Medicinal Applications	Representative Image of Plant
Sesbania grandiflora Pers	Leguminosae	Kathurumurunga	For diabetes, fever, cough, asthma and catarrh, improving memory power, cracked lips, joint swellings and night blindness	 
Solanum trilobatum Linn	Solanaceae	Wel thibbatu	As an emetic and blood purifier, for food poisoning and phlegm-related diseases	

(*Continued*)

TABLE 3.1 (*Continued*)
Herbs of Sri Lanka Prepared in Porridge Form: Scientific Name, Family, Vernacular Name, Traditional Medicinal Applications and Representative Images of Some of These Conventionally Documented Herbs Used in Past and Present Times

Scientific Name	Family	Vernacular Name (Sinhala)	Traditional Medicinal Applications	Representative Image of Plant
Solanum surattense Burm	Solanaceae	Elabatu	As an expectorant, cure for coughs and phlegm-related diseases, improving appetite, for haemorrhoids, liver issues, skin diseases, fever and catarrh	
Syngonium podophyllum	Araceae	Wel kohila	For haemorrhoids, skin diseases and joint swellings	

(*Continued*)

TABLE 3.1 (*Continued*)

Herbs of Sri Lanka Prepared in Porridge Form: Scientific Name, Family, Vernacular Name, Traditional Medicinal Applications and Representative Images of Some of These Conventionally Documented Herbs Used in Past and Present Times

Scientific Name	Family	Vernacular Name (Sinhala)	Traditional Medicinal Applications	Representative Image of Plant
Tragia involucrata Linn	Euphorbiaceae	Wel kahambiliya	For snake bites, skin diseases, dry cough, bronchitis, joint pains and fever	
Vernonia cinerea Linn	Compositae	Monarakudumbiya	For conjunctivitis, rheumatism, appendicitis, encephalitis, skin diseases, prolonged fever, cough and cold	
Vetiveria zizanioides Linn	Graminae	Savandarā	For cholera, headaches, dandruff, heart disease, malaria and meningitis, as a blood purifier	

TABLE 3.2
Selected Scientifically Elucidated Health Benefits (Currently Available) of Traditional Medicinal Herbs Used to Prepare Porridges in Sri Lanka

Scientific Name	Health Benefits Elucidated Through Scientific Studies
Abrus precatorius Linn	Anti-diabetic (Monago and Alumanah 2005; Ediriweera and Ratnasooriya 2009), anti-tuberculosis (Limmatvapirat et al. 2004), anti-plasmodial (Limmatvapirat et al. 2004), and immuno-potentiator (Ramnath et al. 2002)
Abutilon indicum Sweet	Anti-diabetic (Ediriweera and Ratnasooriya 2009; Krisanapun et al. 2009), antioxidant (Chakraborthy and Ghorpade 2010), hepatoprotective (Porchezhiana and Ansarib 2005) and anti-ulcer activity (Weragoda 1980; Dashputre and Naikwade 2011)
Acalypha indica Linn	Wound-healing (Reddy et al. 2002), antimicrobial (Somchit et al. 2010), analgesic, and anti-inflammatory (Rahman et al. 2010)
Achyranthes aspera Linn	Anti-inflammatory (Vetrichelvan and Jegadeesan 2003), anti-diabetic (Akhtar and Iqbal 1991), and antioxidant (Priya et al. 2010)
Aerva lanata Linn	Antioxidant (Hara et al. 2018), anti-proliferative, and apoptotic activity (Anusha et al. 2016; Krisnamoorthi and Elumalai 2018), anti-diabetic (Ediriweera and Ratnasooriya 2009; Akanji et al. 2018), anti-microbial (Devarai et al. 2017), and anti-HIV (Gujjeti and Mamidala 2014)
Alangium salvifolium Linn	Anti-cancer (Zahan et al. 2011), antioxidant (Jain et al. 2010), antimicrobial (Jain et al. 2010), analgesic, and anti-inflammatory (Porchezhian et al. 2001)
Alternanthera sessilis Linn	Antioxidant (Borah et al. 2011; Lee et al. 2014), wound-healing (Jalalpure et al. 2008), anti-inflammatory (Muniandy et al. 2018), and analgesic (Hossain et al. 2014)
Asparagus racemosus Wild	Antioxidant (Lee et al. 2014), anti-ulcer (Goel and Sairam 2002), and starch hydrolase inhibitory activity (Lee et al. 2014)
Astercantha longifolia	Antioxidant (Jayawardena et al. 2015), starch hydrolase inhibitory activity (Jayawardena et al. 2015), and anti-diabetic (Fernando et al. 1991; Ediriweera and Ratnasooriya 2009)
Atlantia ceylanica Linn	Hepatoprotective (Oh et al. 2002), anti-diabetic (Senadheera and Ekanayake 2012), mild antioxidant (Fernando and Soysa 2015), and antibacterial (Munasinghe et al. 2015)
Cardiospermum microcarpum HBK Nov	Antioxidant (Jayanthi et al. 2012), anti-inflammatory (Jayanthi et al. 2009; Jayanthi et al. 2013), and antimicrobial (Weragoda 1980; Reddy et al. 2010)
Cassia auriculata Linn	Antioxidant (Jayawardena et al. 2015), starch hydrolase inhibitory activity (Jayawardena et al. 2015), and anti-diabetic (Pari and Latha 2002; Ediriweera and Ratnasooriya 2009)
Ceiba pentandra Linn	Anti-diabetic (Djomeni et al. 2006a, 2006b), prostaglandin biosynthesis (Noreen et al. 1998), antioxidant (Loganayaki et al. 2013), and antimicrobial (Kubmarawa et al. 2007)

(*Continued*)

TABLE 3.2 (*Continued*)
Selected Scientifically Elucidated Health Benefits (Currently Available) of Traditional Medicinal Herbs Used to Prepare Porridges in Sri Lanka

Scientific Name	Health Benefits Elucidated Through Scientific Studies
Centella asiatica Urb	Gamma-Amino Butyric Acid (GABA)-modulator (Jana et al. 2010; Savage et al. 2018), antioxidant (Lee et al. 2014; Waisundara and Watawana 2014), starch hydrolase inhibitory activity (Lee et al. 2014), anti-diabetic (Ullah et al. 2009), anti-cancer (Ullah et al. 2009; Chauhan et al. 2010), anti-inflammatory (Chatterjee et al. 1992; Bunpo et al. 2004), cardioprotective (Chatterjee et al. 1992; Bunpo et al. 2004), antimicrobial (Ullah et al. 2009; Chauhan et al. 2010), and anti-hypertensive (Ullah et al. 2009; Chauhan et al. 2010)
Clitoria ternatea Linn	Anti-inflammatory (Devi et al. 2003), analgesic (Devi et al. 2003), anti-pyretic (Devi et al. 2003), anti-diabetic, and anti-hyperlipidemic (Daisy et al. 2009)
Dregea volubilis Linn	Anti-inflammatory (Biswas et al. 2009b), antioxidant (Biswas, Haldar and Ghosh 2010), and analgesic (Weragoda 1980; Biswas et al. 2009a)
Eclipta prostrata Linn	Hepatoprotective (Lin et al. 1996), hypolipidemic (Zhao et al. 2015), anti-venom (Pithayanukul et al. 2004), anti-HIV (Tewtrakul et al. 2007), and anti-inflammatory (Arunachalam et al. 2009)
Erythrina variegata Linn	Antioxidant (Kumar et al. 2010), antibacterial (Haque et al. 2006), analgesic (Haque et al. 2006), anti-inflammatory, anti-osetoporotic (Zhang et al. 2007), and anti-cardiovascular effects (Benson et al. 2008)
Evolvulus alsinoides Linn	Anti-helminthic (Dash et al. 2002), adaptogenic (Siripurapu et al. 2005), anti-amnesic (Siripurapu et al. 2005), anti-stress (Siripurapu et al. 2005), and antimicrobial (Dash et al. 2002)
Feronia limonia Linn	Anti-tumor (Saima et al. 2000), antimicrobial (Jayashree and Londonkar 2014), and hepatoprotective (Jain et al. 2011)
Ficus hispida Linn	Hepatoprotective (Weragoda 1980; Mandal et al. 2000), antioxidant (Shanmugarajan et al. 2008), anti-diarrheal (Mandal and Kumar 2002), antinociceptive (2011), anti-neoplastic (Pratumvinit et al. 2009), and cardioprotective (Shanmugarajan et al. 2008)
Ficus religiosa Linn	Anti-diabetic (Ediriweera and Ratnasooriya 2009; Kirana et al. 2009), antioxidant (Kirana et al. 2009), anti-inflammatory (Sreelakshmi et al. 2007), analgesic (Sreelakshmi et al. 2007), and antimicrobial (Rajiv and Sivaraj 2012)
Hemidesmus indicus Linn	Antioxidant (Jayawardena et al. 2015) and starch hydrolase inhibitory activity (Jayawardena et al. 2015)
Lasia spinosa Linn	Antinociceptive, anti-inflammatory, and anti-diarrheal (Weragoda 1980; Deb et al. 2010)
Murraya koenigi Spreng	Antioxidant (Tachibana et al. 2003; Rao et al. 2006), hepatoprotective (Gupta and Singh 2007), anti-cancer (Kok et al. 2012), antimicrobial (Reisch et al. 1994; Pande et al. 2009), anti-leukemia (Chakraborty et al. 1973), anti-trichomonal (Adebajo et al. 2006), and anti-diarrhea (Tachibana et al. 2003)

(*Continued*)

TABLE 3.2 (*Continued*)
Selected Scientifically Elucidated Health Benefits (Currently Available) of Traditional Medicinal Herbs Used to Prepare Porridges in Sri Lanka

Scientific Name	Health Benefits Elucidated Through Scientific Studies
Osbeckia octandra DC Prodr	Hepatoprotective (Thabrew et al. 1987), antioxidant, and anti-diabetic (Perera et al. 2013)
Pandanus tectorius Soland	Anti-hyperlipidemic (Zhang et al. 2013), hepatoprotective (Zhang et al. 2013), antioxidant, and antimicrobial (Andriani et al. 2015)
Phyllanthus debilis	Hepatoprotective (Weragoda 1980; Sane et al. 1995) and antioxidant (Kumaran and Karunkaran 2007)
Sesbania grandiflora Pers	Anti-diabetic (Ediriweera and Ratnasooriya 2009; Kumar et al. 2015), antioxidant (Lee et al. 2014; Waisundara and Watawana 2014), starch hydrolase inhibitory activity (Lee et al. 2014), anti-cancer (Pajaniradje et al. 2014), antimicrobial (China et al. 2012), and anti-helminthic (Bhalke et al. 2010)
Solanum trilobatum Linn	Hepatoprotective (Shahjahan et al. 2004) and antimicrobial (Weragoda 1980; Latha and Kannabiran 2006)
Solanum surattense Burm	Antimicrobial (Weragoda 1980; Liu et al. 2006) and antioxidant (Muruhan et al. 2013)
Syngonium podophyllum	Antioxidant, antibacterial, and cytotoxic effects (Kumar et al. 2014)
Tragia involucrata Linn	Wound-healing (Weragoda 1980; Samy et al. 2006), anti-inflammatory, and analgesic (Gobalakrishnan et al. 2013)
Vernonia cinerea Linn	Anti-cancer (Kuo et al. 2003), anti-malarial (Soma et al. 2017), antioxidant (Goggi and Malpathak 2017), and for cessation of smoking (Wongwiwatthananukit et al. 2009)
Vetiveria zizanioides Linn	Antioxidant (Weragoda 1980; Kim et al. 2005; Luqman et al. 2009)

further in detail in other sub-chapters in the book. Many of these herbs are also consumed in salad form together with rice and curry. However, as noted previously, incorporation into porridge is a more appetizing preparation since rice tends to mask many of the associated bitter flavors and off-notes. Older populations of Sri Lanka, however, usually prefer consuming the herbs as porridge with less mastication and easy digestion. Also, since the herbs are ground up into porridge, the phytonutrients might be better extracted and exposed and thus increase their bioavailability. Although the herbs consumed in porridge form are mostly creepers and small plants, their leaves can indeed be quite thick and even though they are cut into pieces to be made into salads, there is still an amount of breakdown required for the nutrients to be absorbed into the body. Thus, where an easy and rapid intake of phytonutrients is involved, it might be better to consume these herbs in porridge form rather than as salads.

Importantly, the herbs incorporated into porridge need to be obtained from locations that are clean and void of animal excrement, especially when procuring them from forest areas. Sometimes, the leaves of herbs such as *Acalypha indica* can be eaten by domesticated animals (especially cats) as a self-administered medication for stomach upsets. In such instances, the remainder of the plant after animal consumption should not be taken for preparation of porridge. These herbs are also vulnerable to damage by insects and worms since they do not grow to a significant height. Remnants of herbs that have been subjected to such attacks should be carefully removed prior to obtaining the herbs for porridge preparation as well. It is actually of value that most of these herbs are small plants or creepers, since they can be easily ground into a paste or pulp on a grindstone or ground in a kitchen blender, so that the fluid extract can be obtained and added to the rice gruel. In modern times, though, most of these herbs in cleaned and cleansed form can be bought commercially and generally washed in salt water prior to use.

Some of these herbs such as *Centella asiatica*, *Asparagus racemosus*, *Alternanthera sessilis*, *Acalypha indica*, and *Aerva lanata* can be found growing in the wild or in backyards of houses in Sri Lanka. Nevertheless, regardless of their abundance, some of these herbs have been converted into instant herbal porridge mixtures such as those shown in Figure 3.3, where boiling water needs to be added to the sachets in the boxes and the resulting product can be consumed immediately. The emergence of these products is evidence of catering to the high-paced lifestyle of many of the local population, and such instant, ready-to-cook, and ready-to-eat herbal porridges are certainly in demand among Sri Lankans living overseas as well. A typical example of porridge mixtures found on online shopping for Sri Lankans is found at Kapruka.com.

FIGURE 3.3 (a) *Murraya koenigi* and (b) *Asparagus racemosus* instant herbal porridge mixtures, which can be readily obtained from supermarkets in Sri Lanka. These are purchased more by the younger population that leads a demanding lifestyle but nevertheless wish to maintain health and wellness.

There are several herbs listed in Table 3.1 that require further comment. *C. asiatica* is a very popular herb essentially grown in almost all Sri Lankan backyards. It is a very robust plant and has a very rapid growth rate, thus requiring very little caretaking and nurturing to have a significant yield. *C. asiatica*-based herbal porridge is most popular nowadays, with *Murraya koenigi* coming a close second (Figure 3.3). These two types of herbal porridges might be consumed by locals at least once a week for breakfast. Consuming other types of herbal porridge is considered quite 'fashionable' as well, especially those that are made from the more unique herbs such as *Dregea volubilis*, *Asteracantha longifolia*, and *Solanum surattense*. Although the scientific evidence in support of these listed herbs is increasing, the proof in support of their consumption as porridge is comparatively less. In yet another study by Senadheera et al. (2014), *Asparagus racemosus*-, *Scoparia dulcis*-, and *Hemidesmus indicus*-based porridges were tested for their anti-diabetic properties in diabetic Wistar rats, and it was encouraging to see these porridges demonstrating weight loss, hypoglycaemic, and hypolipidemic effects.

CONCLUDING REMARKS

Considered to be healthy and packed with different favorable effects in terms of health, the herbal porridge recipe of Sri Lanka and the habit of consumption appears to be timeless. It is a popular breakfast option among many Sri Lankans owing to its densely packed nutrients and ease of preparation as a food item. The herbs incorporated into the porridge have been scientifically studied for preventing various ailments such as diabetes, cancer, heart disease, and even joint issues such as arthritis. Although many of these herbs are also eaten as salads with rice and curry, the Sri Lankans see their consumption with rice porridge as more palatable and a quick way of digestion. The development of instant herbal porridge mixtures is a growing trend among food manufacturers of Sri Lanka, to meet the demands of urbanized consumers of the country. The taste of these instant herbal porridges may be quite different from those that are prepared fresh, but nevertheless, there appears to be a demand for such products, especially among the younger generation who are more health conscious despite their fast-paced lifestyle and corporate careers. Overall, as a popular and healthy food habit in Sri Lanka, even though the individual herbs have been studied there is still a lack of scientific evidence in support of consuming herbal porridges and further attention is warranted to elucidate its beneficial effects. Through increased scientific evidence, the habit can be further cultivated among the locals and can even be promoted globally as a healthier, more flavorful, and pleasant option of consuming herbs, especially those that are considered as leafy greens.

CENTELLA ASIATICA

C. asiatica is one of the most frequently consumed herbal leaves/leafy vegetables of Sri Lanka and deservedly is allocated a separate chapter. It is typically added into a salad or to rice gruel in porridge form. As mentioned in the chapter on herbal porridges, *C. asiatica*-based rice porridge is a very popular breakfast item among locals; in fact, it is often consumed for overall health and wellness purposes. The porridge form of the herb was once offered to pre-school children in Sri Lanka to combat nutritional deficiencies (Cox et al. 1993; Chandrika and Kumara 2015). *C. asiatica* appears to be one of the 25 top-selling herbs in the United States as well (Siddiqui et al. 2011). In addition to its nutritional and pharmaceutical value, *C. asiatica* is also known as a functional food and a remedy for many diseases in the folk medicinal systems around the world. It is considered as a 'panacea' in China, India, Africa, the Philippines, and Madagascar (Bylka et al. 2013). The vernacular names used to refer to *C. asiatica* are summarized in Table 3.3. *C. asiatica* is a well-researched herb and there are several published reviews highlighting its

TABLE 3.3
Vernacular Names Used for *C. asiatica*

Country	Language/Region	Vernacular Name
Sri Lanka	Sinhala	Gotu kola
India	Hindi	Bemgsag, Brahma-Manduki, Gotukola, Khulakhudi, Mandookaparni
	Malayalam	Kodagam, Kodangal, Kutakm, Kutannal, Muthal, Muttil, Muyalchevi
	Telugu	Bekaparnamu, Bokkudu, Saraswataku, Mandukbrahmmi, Saraswati plant
	Marathi	Karinga, Karivana
	Tripura	Thankuni, Thunimankuni
	Assam	Manimuni
	Bihar	Chokiora
	Oriya	Thalkudi
	Urdu	Brahmi
	Sanskrit	Bhekaparni, Bheki, Brahmamanduki, Darduchhada, Divya, Mahaushadhi, Mandukaprnika, Manduki, Mutthil, Supriya, Tvasthi
	Kanarese	Brahmisoppu, Urage, Vandelagaillikiwigidda, Vondelaga
	Gujarati	Barmi, Moti Brahmi
	Tamil	Babassa, Vallarai
	Bengal	Thankuni, Tholkuri
	Deccan	Vallarai
	Meghalaya	Bat-maina
USA	English	Indian Pennywort, Marsh Pennywort
	Hawaii	Pohe Kula
Oceania	Cook Islands	Kapukapu
	Fiji	Totodro
Malaysia	Malay	Pengaga
China		Fo-ti-tieng, Chi-hsueuh-ts'ao
Nepal		Ghod tapre
Tahiti		Tohetupou
Samoa, Tonga		Tono

Source: Singh, S. et al., *Int. J. Pharm. Sci. Rev. Res.*, 4, 9–17, 2010.

therapeutic effects and bioactive compounds (Arora et al. 2002; James and Dubrey 2009; Kohil et al. 2010; Bhavna and Jyoti 2011; Orhan 2012; Chong and Aziz 2013; Bylka et al. 2014; Naveen et al. 2017; Prakash et al. 2017; Sabaragamuwa et al. 2018). Efficacy of the plant extracts for various therapeutic effects have been examined in a few human clinical trials (Pointel et al. 1987; Bradwejn et al. 2000; Brinkhaus et al. 2000; Upadhyay et al. 2002; Jana et al. 2010). *C. asiatica* has appeared in almost all the traditional herbal medicinal systems of the world as a highly effective agent to enhance memory power (Arora et al. 2002; James and Dubrey 2009; Kohil et al. 2010; Bhavna and Jyoti 2011; Orhan 2012). Although considered as a weed, the herb is cultured for horticultural purposes and is found in abundance in most backyards in Sri Lanka.

Origins, Morphology, and Growth

C. asiatica belongs to the Umbelliferae/Apiaceae family. The taxonomic classification of *C. asiatica* is shown in Table 3.4. It is slender and stoloniferous with a weak aroma and is also considered a perennial weed. Figure 3.4 displays its leaves and root structure. It is especially abundant in the swampy areas of Sri Lanka and India, up to an altitude of approximately 700 m (Arora et al. 2002). The herb is a prostrate creeper, which can attain a height up to 15 cm (Singh et al. 2010). The stem of the plant is glabrous, striated with rooting at the nodes. *C. asiatica* flourishes extensively in shady, marshy, damp, and wet places such as paddy fields and river banks, forming a dense green carpet (Singh et al. 2010). Sandy loam (60% sand) is found to be the most fertile soil for its regeneration (Anjana and Pramod 2009). The leaves, 1–3 from each node of stems, are long petioled, 2–6 cm in length and 1.5–5 cm wide, in orbicular-renniform with a sheathing leaf base, crenate margins, and glabrous on both sides (Singh et al. 2010). Flowers are in fascicled umbels, each umbel consisting of 3–4 white to purple or pink flowers (Singh et al. 2010). Flowering occurs in the month of April–June (Singh et al. 2010). Fruits are borne throughout the growing season in approx. 2″-long, oblong, globular shapes and strongly thickened pericarp. Seeds have pedulous embryos, which are laterally compressed

TABLE 3.4
Taxonomic Classification of *C. asiatica*

Kingdom	Plantae
Sub-kingdom	Viridiplantae
Division	Tracheophyta
Sub-division	Spermatophytina
Class	Magnoliopsida
Order	Apiales
Family	Apiaceae
Genus	*Centella* L.
Species	*Centella asiatica* (L.) Urb.

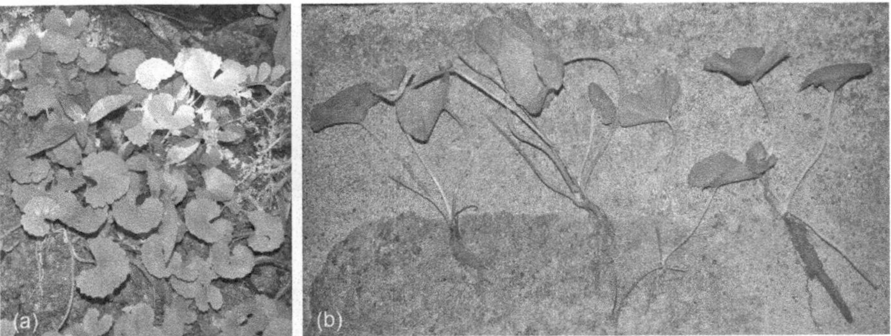

FIGURE 3.4 Leaves and root structure of *C. asiatica*.

(Singh et al. 2010). There are several morphotypes of *C. asiatica* that have been reported from different countries of the world (Chandrika and Kumara 2015). These primarily vary in terms of the leaves, leaf length, thickness of leaf, texture, leaf color, leaf margin, pigments, and abundance of flowers (Chandrika and Kumara 2015).

Despite its importance, no serious efforts have been devoted to planning and organizing the commercial propagation and cultivation of *C. asiatica* around the world. The spontaneous collection of the plant from natural sources and over-exploitation are now widespread in response to high market demand (Siddiqui et al. 2011). In this study the conjunctive use of a compost tea and an inorganic fertilizer on the growth, yield, and terpenoid content of *Centella asiatica* (L.) urban was evaluated. It was discovered that compost tea and inorganic fertilizer applied at half the recommended concentration resulted in a significant enhancement of vegetative growth, yield, and antioxidant content of the plants.

Traditional Medicinal Applications

A summary of the common uses of *C. asiatica* among the traditional medicinal systems around the world is shown in Figure 3.5. In Sri Lanka in particular, the herbal extract has been used for the treatment of catarrh, asthma, eye infections, spider bites, snake bites, haemorrhoids, headaches, sore throat, cracked lips, and improve memory. In light of the increasing concerns of Alzheimer's disease with the ageing population, *C. asiatica* for improving memory takes on even more importance. In modern times, the extract of the herb is sometimes mixed with ginger and black pepper for the treatment of coughs. In the Indian traditional medicinal system, it is a popular 'Rasayana' herb and is used for the treatment of epilepsy, schizophrenia, and cognitive dysfunction (Arora et al. 2002). It has also been used for the treatment of renal stones, leprosy and skin diseases, anorexia, and asthma in India (Arora et al. 2002). The wound-healing properties of the herb have been used in the traditional medicinal systems of the Malay peninsula, Java, and Madagascar, whereas psychotropic applications of the herbal extract have been recorded in China (Bhavna and Jyoti 2011).

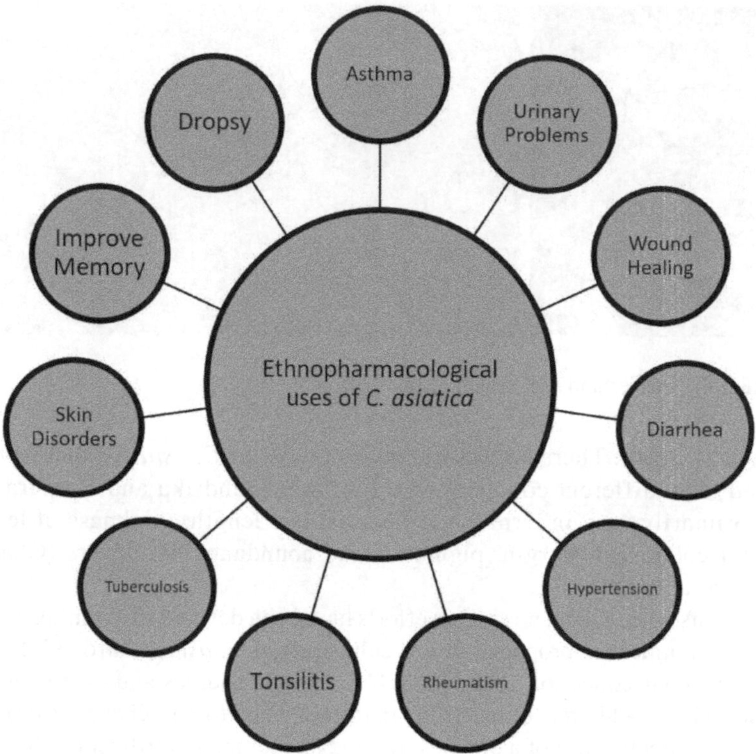

FIGURE 3.5 Common uses of *C. asiatica* among most of the traditional medicinal systems around the world.

C. asiatica has been used for dermatological conditions, to improve small wounds, scratches, and burns; for hypertrophic wound healing; and as an anti-inflammatory agent, particularly against eczema (Bylka et al. 2013). This accounts for its modern-day usage of incorporation into cosmetics. *C. asiatica* is commonly eaten fresh as a vegetable, especially by the local Malay and Javanese populations (Huda-Faujan et al. 2007). The salads are eaten together with the main meal and can act as an appetizer. It is also cooked as a part of a soup or as a main vegetable. In Thailand and India, it is used as a tonic drink and juice (Punturee et al. 2004). *C. asiatica* herbal tea is made by adding a cup of boiled water over either dried or fresh *C. asiatica* plant tissues, letting it brew a few minutes before drinking (Hashim et al. 2011).

Bioactive Compounds of *C. asiatica* and Their Therapeutic Effects

The content of the pharmacologically active components in *C. asiatica* is typically influenced both genetically and environmentally, depending on cultivation conditions (Siddiqui et al. 2011). Cultivation conditions could be optimized further to maximize the synthesis of pharmacologically active constituents, although the limits of such biosynthesis will eventually be determined by the genetic composition of the plant and reach a maximum threshold (Samuelsson 1992). Numerous preparations of *C. asiatica* in various pharmaceutical formulations have been recommended for several indications including neurological disorders. Thus, many researchers have focused on the neuroprotective effects of *C. asiatica* to provide evidence in support of its traditional use. A key target of most pharmacotherapies that focuses on the treatment of anxiety disorders is gamma-aminobutyric acid (GABA), which is an inhibitory neurotransmitter within the central nervous system. Specifically, anxiety disorders are chronic and functionally these are disabling conditions with high psychological stress, characterized by cognitive symptoms of excessive worry (Savage et al. 2018). Awad et al. (2007) reported that *C. asiatica* significantly increased glutamic acid decarboxylase activity *in vitro* by more than 50% ($p < 0.001$) in male Sprague–Dawley rat brain assays. In another *in vitro* study of GABA-A subtype receptor modulation by Hamid et al. (2016), Asiatic acid (Figure 3.6) was found to be a negative modulator of GABA-induced currents for α1 β2 γ2L, α2 β2 γ2L, and α5 β3 γ2L receptors in Xenopus oocyte tissue. Further, Chatterjee et al. (1992) performed an *ex vivo* study using Charles Foster rats and *C. asiatica* extract increased whole brain levels of GABA. A human study examined the effects of an acute dose of *C. asiatica* extract (12 g) on measures of stress and anxiety, including acoustic startle response, against a placebo comparator in 40 healthy adults (Bradwejn et al. 2000). Results indicated a significant attenuation of acoustic startle response in the treatment group. Jana et al. (2010) conducted a clinical investigation where 500 mg of the *C. asiatica* extract was administered twice daily for 2 months on 33 patients (18 males and 15 females) with generalized anxiety disorder (Jana et al. 2010). The study found that *C. asiatica* not only significantly ($p < 0.01$) attenuated anxiety-related disorders but also significantly ($p < 0.01$) reduced stress phenomena and its correlated depression.

Ursane-type pentacyclic triterpenoids known as centelloids, mainly, asiaticoside, madecasosside (brahminoside), asiatic acid (Figure 3.6), and madecassic acid (brahmic acid) are shown in Figure 3.7 and are considered the most important bioactive constituents

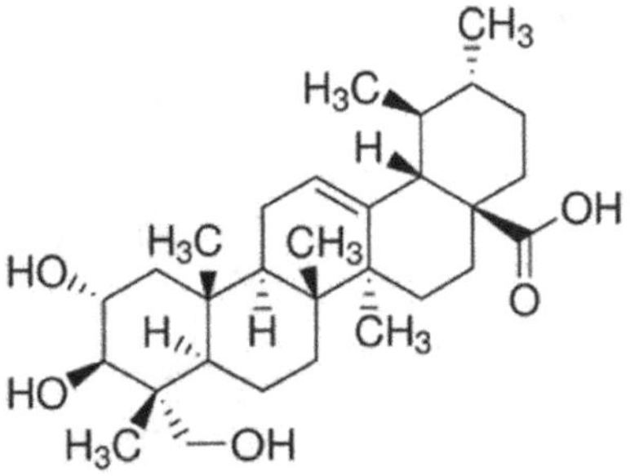

FIGURE 3.6 Asiatic acid was found to be a negative modulator of GABA-induced currents for α1 β2 γ2L, α2 β2 γ2L, and α5 β3 γ2L receptors in Xenopus oocyte tissue. (From Hamid, K. et al., *Chem. Biol. Drug Des.*, 88, 386–397, 2016.)

isolated from *C. asiatica* (Orhan 2012; Bylka et al. 2014). Several flavonoids such as quercetin, kaempferol, patuletin, rutin, apigenin, castilliferol, castillicetin, and myricetin have been reported in the whole plant (Kuroda et al. 2001; Matsuda et al. 2001; Subban et al. 2008). *C. asiatica* is also rich in Vitamin C, Vitamin B1, Vitamin B2, niacin, carotene, and Vitamin A (Pal and Pal 2016). The presence of polysaccharides (Wang et al.

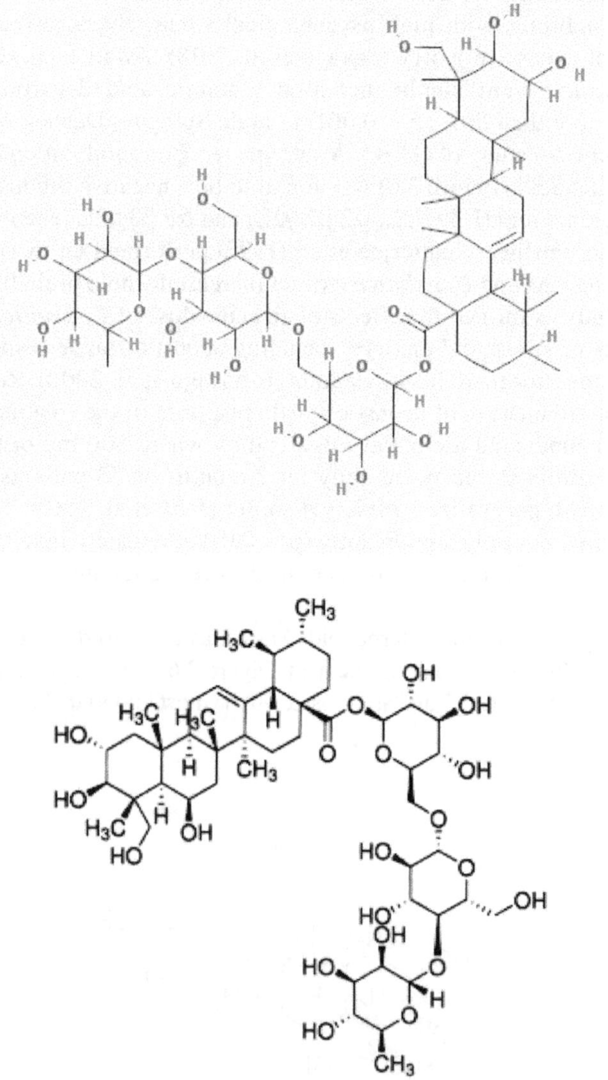

(a)

(b)

FIGURE 3.7 Other chemical constituents found in *C. asiatica*: (a) asiaticoside, (b) madecassoside. *(Continued)*

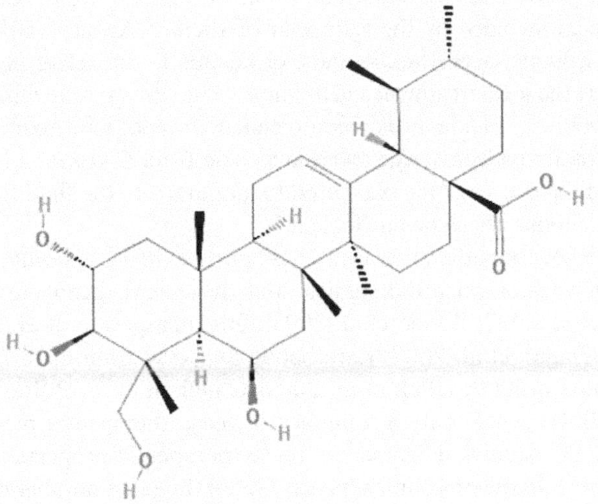

(c)

FIGURE 3.7 (Continued) Other chemical constituents found in *C. asiatica*: (c) madecassic acid (brahmic acid).

2004), polyacetylenes (Govindan et al. 2007; Siddiqui et al. 2011), sterols (Srivastava and Shukla 1996; Rumalla et al. 2010; Sondhi et al. 2010), and phenolic acids (Suntornsuk and Anurukvorakun 2005; Yoshida et al. 2005; Rumalla et al. 2010) have been also identified in the whole extracts of the species.

Lin et al. (2017) examined the combination of asiatic acid, madecassic, madecassoside, quercetin, and isoquercetin originating from *C. asiatica* as a substitute for nerve growth factor. In this study, 16 different combinations of these compounds were investigated for their neuroprotective activities on cultured pheochromocytoma PC12 cells, which are commonly used to evaluate neuronal differentiation in responses to various stimuli. The experimental results were analyzed through statistical modelling and the optimal combination of the compounds, which promotes the highest neurite elongation in PC12 cells, was verified further. It was observed that the individual compounds had minimal differentiation effects, whereas the optimal drug combinations promoted differentiation significantly.

C. asiatica extract (International Nomenclature of Cosmetic Ingredients, INCI) is used also as an ingredient of cosmetics (Bylka et al. 2013; Bylka et al. 2014). The extracts are available in tablet form as well, and information on the medicinal products suggests that all extracts – titrated extract of *C. asiatica* (TECA), total triterpenoid fraction of *C. asiatica* (TTFCA), total triterpenoid fraction (TTF), as well as *C. asiatica* total triterpenoid fraction (CATTF) and *estratto titolato* di-titrated extract of *C. asiatica* (ETCA) – are different acronyms of the same extract, contained in the commercial products, Madecassol®, Centellase® or Blastoestimulina® (Bylka et al. 2014). These extracts include 40% of asiaticoside and a 60% mixture of asiatic and madecassic acids (Brinkhaus et al. 2002; European Medicines Agency 2018). One to two tablets (10 mg) three times a day for adults and a half of this dose for children under 3 years of age are recommended by the European Medicines Agency (2018) in the case of non-healing wounds, hypertrophic scars, or keloids in the active phase. For external use, to support the local treatment and to improve the granulation phase of non-healing ulcers and wounds, 1% cream is recommended. In addition, many commercial formulations contain madecassoside and asiaticoside from *C. asiatica* in different ratios, depending on the source of the plant used to manufacture the final formulation. Some of these applications are shown in Table 3.5.

There are several other therapeutic properties of *C. asiatica* that have been elucidated in various scientific studies, and these have been previously reviewed (Chatterjee et al. 1992; Bunpo et al. 2004; Gnanapragasam et al. 2004; Guo et al. 2004; Huda-Faujan et al. 2007; Karalliadde and Gawarammana 2008; Ediriweera and Ratnasooriya 2009; Ullah et al. 2009; Chauhan et al. 2010; Waisundara and Watawana 2014); hence, only a summary of these therapeutic properties is shown in Table 3.5. The studies in support of these therapeutic properties include *in vitro* and *in vivo* work in the pre-clinical phase. Given the large number of bioactive compounds that are present throughout the roots and leaves of the plant, it is quite evident that *C. asiatica* offers many routes to preventing and remediating many incurable diseases.

TABLE 3.5
Product Range of Extracts from *C. asiatica* Indicating the Specific Chemical Composition and Treatment

Type of Extract/ Chemical Compound	Chemical Composition	Applications
Asiatic acid	>95% Asiatic acid	Anti-ageing cosmetics, application after laser therapy, cosmeceuticals
Titrated extract of *Centella asiatica*	55%–66% Genins, 34%–44% Asiaticoside, >40% Genins, >36% Asiaticoside	Anti-cellulite, slimming products, breast creams, stretch marks, scarred skin, anti-ageing cosmetics, moisturizing care
Heteroside	>55% Madecassoside, >14% Asiaticoside	Slow release effect, anti-ageing cosmetics, for moisturizing night creams
Asiaticoside	>95% Asiaticoside	Anti-inflammatory, against irritated and reddened skin, anti-allergic
Genins	>25% Asiatic acid, >60% Madecassic acid	Natural antibiotic, antibacterial properties, for anti-acne products, intimate hygiene

Source: James, J.T. and Dubrey, I.A., *Molecules*, 14, 3922–3941, 2009.

Concluding Remarks

C. asiatica remains an invaluable herb in almost all the traditional medicinal systems of the world. Its usage for various ailments appears to be common despite the global dispersion of the traditional medicinal systems. This commonality could be due to the bioactive compounds present in the plants' extract, since their mechanisms of action would eventually lead to preventing or remediating the same disease conditions. The growing popularity of *C. asiatica* in Sri Lanka and its increased consumption in porridge form as a breakfast item offer encouraging signs. As mentioned in the chapter on herbal porridges, the herb has been converted to a convenient packet form and made available to the urban consumer market, especially supermarkets of Sri Lanka. With the introduction of such products, the interest of Sri Lankan consumers in *C. asiatica*-based items keeps increasing and the herb remains a vital element to be cultivated in the households of the country.

MURRAYA KOENIGII (L.) SPRENG

M. koenigii is known as 'karapincha' in Sinhala or curry leaves in English. Since these leaves are frequently used in curries they are referred as 'curry leaves', although in most of the Indian sub-continental languages, the vernacular names translate to 'sweet neem leaves'. The species name of the plant is used to venerate the botanist Johann König, while the genus *Murraya* commemorates Swedish physician and botanist Johan Andreas Murray. The taxonomic classification of *M. koenigii* is shown in Table 3.6, while the vernacular names used to refer to this plant are shown in Table 3.7. At present, apart from its food and medicinal value, *M. koenigii* serves as an important component in cosmetics such as lotions and massage oils, in the soap-making industry, in the making of diffusers, potpourri, air-fresheners, perfume oils, aromatherapy products, bath oils, and incense (Rajendran et al. 2014). In Sri Lanka as well as India, in the absence of *Ocimum tenuiflorum* leaves (tulsi), *M. koenigii* leaves are used for Hindu and other religious rites and rituals.

TABLE 3.6
Taxonomic Classification of *M. koenigii*

Kingdom	Plantae
Sub-kingdom	Tracheobionta
Division	Magnoliophyta
Class	Magnoliopsida
Sub-class	Rosidae
Order	Sapindales
Family	Rutaceae
Genus	*Murraya*
Species	*koenigii*

TABLE 3.7
Vernacular Names Used for *M. koenigii*

Language/Region/Country	Name
Bahasa Indonesia	Daun kari
Bengali	Barsunga
Burmese	Pindosine, Pyim daw thein
Danish	Karrry bald
Dutch	Kerriebladeren
English	Curry leaves
French	Feuilles de cari, Feuilles de cury
German	Curryblatter
Gujarati	Mitho limdo
Hindi	Meetha neem, Karipatta, Kathnim, Bursunga
Italian	Fogli de Cari
Kannada	Karibevu
Malayalam	Kariveppilei, Kareapela
Marathi	Karipat, Karhi patta, Karhinimb, Jhirang
Oriya	Bansago
Sanskrit	Girinimba, Suravi
Sinhala	Karapincha
Spanish	Hoja
Tamil	Karivempu, Karuveppilei, Karivepila
Telugu	Karepaku, Karuvepaku

Source: Gahlawat, D.J. et al., *J. Pharmacogn. Phytochem.*, 3, 109–119, 2014.

M. koenigii is an important medicinal plant of Sri Lanka as well as the Indian sub-continent. It is grown and found in almost every household in the region for its medicinal value, and the aromatic leaves are an important ingredient for many of the region's cuisines. The plant is considered as an essential seasoning and flavoring agent and is added into almost every culinary preparation. An herbal porridge made from *M. koenigii* is a popular breakfast item in Sri Lanka. The dried form of the plant is also available in the supermarkets of Sri Lanka as an herbal tea. In culinary preparations, the *M. koenigii* leaves are usually fried along with vegetable oil, mustard seeds, and chopped onions in the first stage of the preparation to release the aroma of the leaves into the oil. The leaves are also available in dried form in supermarkets and from street vendors of Sri Lanka, but they contain less aroma and are largely inferior to the fresh form. In Sri Lankan households, for preservation purposes, the *M. koenigii* leaves are removed from the stems, washed, and patted dry, then placed on a large plate with a sieve or mesh cover and dried out in the sun for 2–3 days. Fresh leaves are also stored in refrigerators, although in this form, the intensity of the aroma decreases over time.

Origins, Morphology, and Growth

M. koenigii is a deciduous shrub or small tree reaching up to 6 m in height. Most parts of the plant are covered with fine down and have a pungent odor (Handral et al. 2012). The plant has a short trunk with a diameter of 15–40 mm, which is smooth, greyish or brown, and has a dense shady crown (Gahlawat et al. 2014). The main stem is dark green to brownish in color (Gahlawat et al. 2014). The leaves are bipinnately compound, 15–30 cm in length, each bearing 11–25 leaflets alternating on rachis, 2.5–3.5 cm in length, ovate and lanceolate with an oblique base (Gahlawat et al. 2014). The leaf margins are irregularly serrated and the petiole is 2–3 mm in length (Gahlawat et al. 2014). Inflorescence is terminal cymes, with each bearing 60–90 flowers (Gahlawat et al. 2014). Each flower is bisexual, white, funnel shaped, sweetly scented, stalked, complete, ebracteate, and regular with an average diameter of a fully opened flower being 1.12 cm (Kumar et al. 1999). The calyx is deeply lobed and pubescent with five clefts (Kumar et al. 1999; Gahlawat et al. 2014). The petals of the flowers are five with free, whitish, glabrous dotted glands (Gahlawat et al. 2014). Fruits occur in close clusters and are small ovoid or subglobose, glandular, with thin pericarp enclosing one or two seeds, which are spinach green in color (Iyer and Mani 1990). Fruits are 2.5 cm long and 0.3 cm in diameter wrinkled with glands and turn purplish black after ripening; they are edible and yield 0.76% of a yellow volatile oil. The individual seed is 11 mm long, 8 mm in diameter, and weighs up to 445 mg (Nutan et al. 1998).

An aerial view of *M. koenigii* leaves and branching is shown in Figure 3.8. *M. koenigii* is found in abundance in tropical Asia such as the foothills of the

FIGURE 3.8 Aerial view of *M. koenigii* leaves and their branching structure.

Himalayas of India, Sri Lanka, Myanmar, and Indonesia. In India, the plant can be found in abundance in Sikkim, Garhwal, Bengal, Assam, Western Ghats, and Cochin (Kumar et al. 2013). It naturally reproduces by the means of seeds, which germinate freely under partial shade, although stem cuttings are the part of the plant used by residents of the Indian sub-continent for propagation purposes. It is also available in other parts of Asia in areas of moist forests such as Guangdong, South Hainan, South Yunnan, Bhutan, Laos, Nepal, Pakistan, Thailand, and Vietnam (Kumar et al. 2013). Brought by south Indian immigrants, *M. koenigii* reached Malaysia, South Africa, and Reunion Island. It is rarely found outside the Indian subcontinent (Jain et al. 2012).

Traditional Medicinal Applications

M. koenigii has been used for centuries in the traditional medicinal systems of the Indian subcontinent (Bhandari 2012). A screening of available literature on *M. koenigii* indicates that it is a common remedy among various ethnic groups and Ayurvedic practitioners for treatment of a diversity of ailments (Bhandari 2012; Jain et al. 2012; Kumar et al. 2013). A summary of the ethnobotanical uses of various parts of *M. koenigii* is shown in Figure 3.9. In the traditional medicinal system of Sri Lanka, the leaf extract has been used for snake and insect bites; improving appetite; preventing liver diseases, worm infections, skin infections, diabetes, and weight loss; as an emetic and improving digestion; and to help prevent untimely growth of grey hairs. In contrast with many of the other herbs discussed in this book, *M. koenigii* leaves are not typically consumed in Sri Lanka on their own in salad form. Typically, they may be ground into a pulp and added together with rice and curry as a chutney.

The leaves of *M. koenigii* have a bitter taste, but the traditional medicinal pharmacopoeia of Sri Lanka, especially *ola leaf* records, indicate that they have been used in both oral and topical form primarily for the purposes of imparting anthelmintic and analgesic action. They are used against haemorrhoids and to reduce body heat, thirst, inflammation, and itching (Bhandari et al. 2012; Yapa 2015). Ancient records indicate that ailments of leukoderma and blood disorders have been controlled using *M. koenigii* leaves (Yapa et al. 2015). Many women in Sri Lanka who are in their first trimester of pregnancy often use *M. koenigii* leaves to relieve any morning sickness and nausea (Yapa 2015). Additionally, ground-up *M. koenigii* leaves are directly applied in paste form onto skin burns, bruises, and skin eruptions, and the paste is sometimes left overnight for more effective results.

Bioactive Compounds in *M. koenigii* and Their Therapeutic Properties

M. koenigii has undergone an extensive study of its chemical constituents and various bioactivities in *in vitro* and *in vivo* studies. Some of the bioactive compounds in *M. koenigii* are presented in Table 3.8 along with their biological activities. The chemical structures of mahanimbine, murrayanine, murrayazoline, isomahanine, koenoline, mukolidine, mukoline, mukonal, and girinimbine, which are the important constituents demonstrating bioactive properties, are shown in Figure 3.10 (Gahlawat et al. 2014).

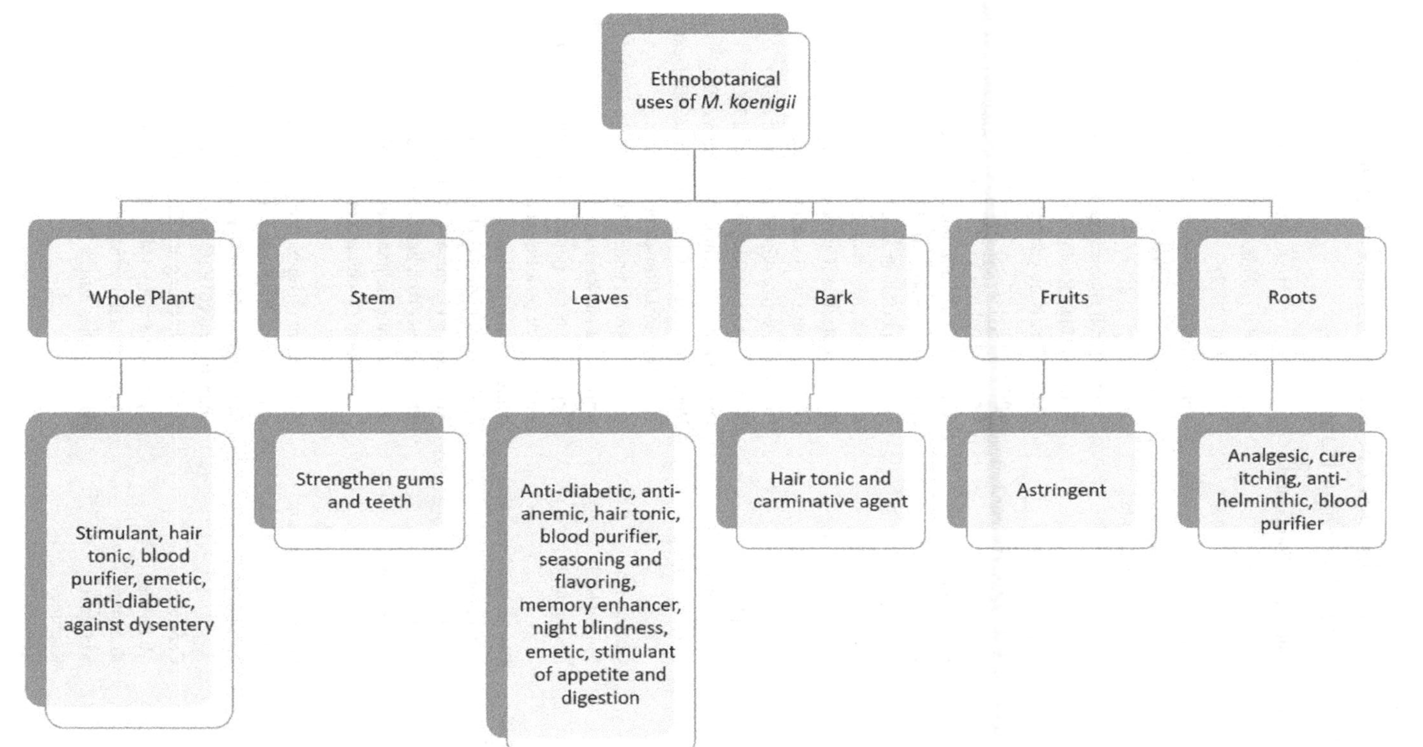

FIGURE 3.9 A summary of the ethnobotanical uses of various parts of the *M. koenigii* plant from all traditional medicinal systems around the world.

TABLE 3.8
Biological Activities of Different Chemical Constituents Identified from Different Parts of the *M. koenigii* Plant

Part of the Plant	Bioactive Compound	Therapeutic Property
Stem bark	Girinimbine	Antibacterial (Kumar et al. 2010) and anti-cancer (Kok et al. 2012)
	Murrayanine	Antibacterial (Kumar et al. 2010)
	Marmesin-1′-O-beta-D-galactopyranoside	Antibacterial (Kumar et al. 2010)
	Mahanine	Antibacterial (Ramsewak et al. 1999)
	Murrayacine	Antimicrobial (Rastogi and Mehrotra 1980)
	Mukoeic acid	Antioxidant (Chowdhury and Debi 1969)
	Murrayazolinine	Anti-leukemia (Chakraborty et al. 1973)
	Girinimbilol	Anti-trichomonal (Adebajo et al. 2006)
Leaves	Koenimbine	Antioxidant and anti-diarrhea (Tachibana et al. 2003)
	Koenine	Antioxidant (Rao et al. 2006)
	Koenigine	Antioxidant (Rao et al. 2006)
	Mahanimbine	Antioxidant (Rao et al. 2006)
	Murrayazolidine	Hepatoprotective (Gupta and Singh 2007)
	Murrayazoline	Hepatoprotective (Gupta and Singh 2007)
	Girinimbine	Hepatoprotective (Gupta and Singh 2007)
	Tocopherol	Hepatoprotective (Gupta and Singh 2007)
	Isomahanimbine	Hepatoprotective (Gupta and Singh 2007)
	Mahanimbine	Hepatoprotective (Gupta and Singh 2007) and antimicrobial (Reisch et al. 1994)
	Gurjunene	Antimicrobial (Pande et al. 2009)
	Murrayanol	Antimicrobial (Reisch et al. 1994)
	Mahanine	Hepatoprotective (Gupta and Singh 2007) and antioxidant (Adesina et al. 1988)
	Bismurrayafoline-E	Antioxidant (Adesina et al. 1988)
	Euchrestine	Antioxidant (Adesina et al. 1988)
	Bismahanine	Antioxidant (Tachibana et al. 2001)
	Bispyrafoline	Antioxidant (Tachibana et al. 2001)
	Isomahanine	Antioxidant (Tachibana et al. 2001)
	O-methyl murrayamine-A	Antioxidant (Tachibana et al. 2001)
	O-methyl mahanine	Antioxidant (Tachibana et al. 2001)
	Lutein	Antioxidant (Tachibana et al. 2003)
	Tocopherol	Antioxidant (Tachibana et al. 2003) and hepatoprotective (Gupta and Singh 2007)
	Carotene	Antioxidant (Song and Chin 1991)
Root	Mukoline	Cytotoxic activity (Manfred et al. 1985)
Seed	Koenoline	Cytotoxic activity (Manfred et al. 1985)
	Kurryam	Anti-diarrhea effects (Mandal et al. 2010)
	Koenine	
	Koenimbine	

Together with the functional properties highlighted in Table 3.8, several studies indicate other therapeutic properties of *M. koenigii*. The hypocholesterolemic activity was checked in aged mice by Tembhurne and Sakarkar (2010), using a crude ethanolic extract of plant leaves of *M. Koenigii* and confirmed by observing a decrease in blood cholesterol levels in a dose-dependent manner. Anti-ulcer activity of both aqueous and ether extracts of *M. koenigii* was studied by Shirwaikar et al. (2006) in a reserpine-induced gastric ulcer model in albino rats. Extracts were effective in gastric ulceration and found to be about as protective as ranitidine.

The *M. koenigii* leaves exhibited hypoglycaemic effects with increased hepatic glycogen content due to increased glycogenesis and decreased glycogenolysis and gluconeogenesis (Khan et al. 1997). An aqueous extract of the leaves of *M. koenigii* exhibited antioxidant activity and superior free radical scavenging ability in a

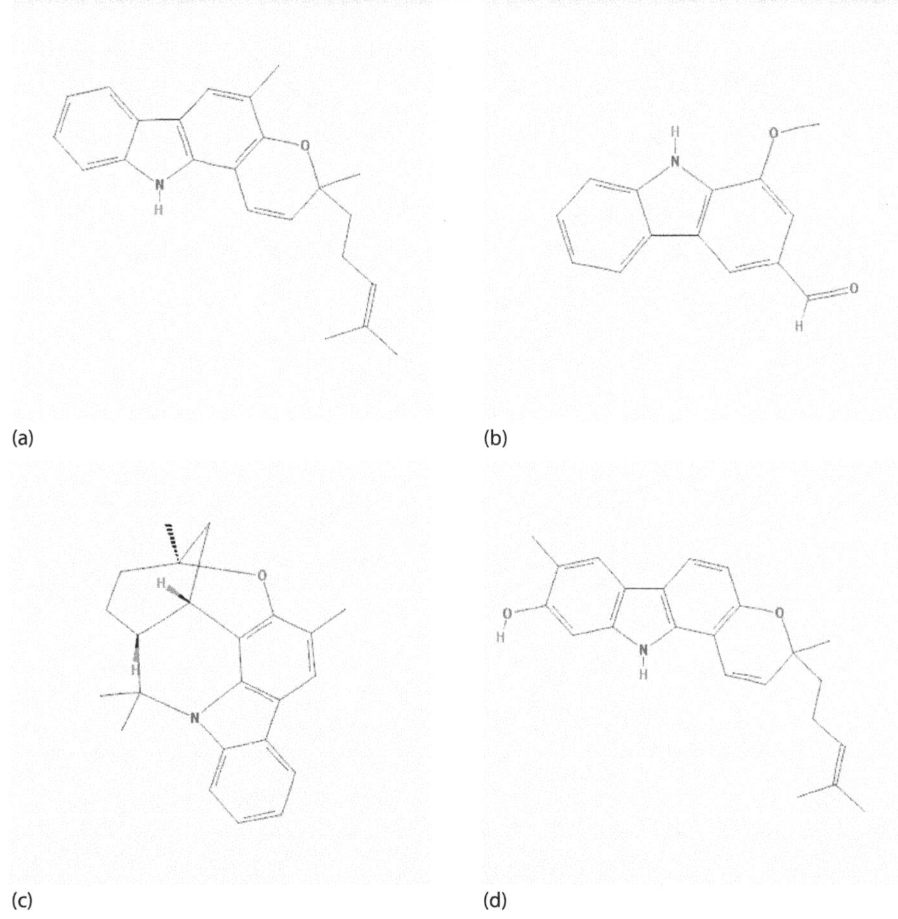

FIGURE 3.10 Some important chemical constituents from *M. koenigii* that have demonstrated bioactive properties: (a) ahanimbine, (b) murrayanine, (c) murrayazoline, (d) isomahanine. *(Continued)*

(e)

(f)

(g)

(h)

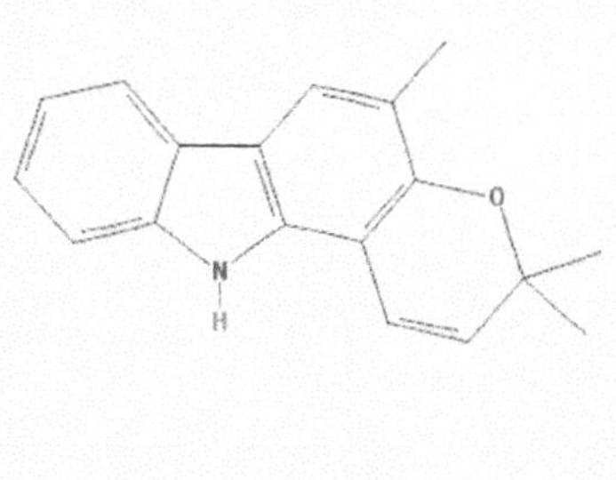

(i)

FIGURE 3.10 (Continued) Some important chemical constituents from *M. koenigii* that have demonstrated bioactive properties: (e) koenoline, (f) mukolidine, (g) mukoline, (h) mukonal, and (i) girinimbine.

hyperglycaemia-induced oxidative stress cell-line model in a study by Choo et al. (2013). In a similar study by Waisundara and Lee (2014), aqueous leaf extracts had high oxygen radical absorbance capacity (ORAC) values and superior free radical scavenging abilities and increased the superoxide dismutase (SOD) activity in hyperglycaemia-induced human umbilical vein endothelial cells (HUVECs). Waisundara (2018) assessed the stability of the antioxidant and starch hydrolase inhibitory properties of *M. koenigii* aqueous leaf extract in an *in vitro* model of gastric and duodenal digestion to assess whether the antioxidant activity and starch hydrolase inhibitory activity of the leaf extract had either been maintained during the digestion phases or significantly increased ($p < 0.05$). They concluded that *M. koenigii* leaf extract was resistant to gastric and duodenal digestion, thus supporting its functional effects in terms of antioxidant and starch hydrolase inhibitory properties.

The fruit of *M. koenigii* is less well known for traditional medicinal applications. The pulp of the fruits generally contains 64.9% moisture, 9.8% total sugar, 9.6% reducing sugar, and negligible amounts of tannin and acids, besides containing 13.4% of vitamin C (Bonde et al. 2011). The pulp of the fruit also contains trace amounts of minerals, ~2% phosphorus, 0.1% potassium, 0.8% calcium, 0.2% magnesium, 0.01% iron, and a remarkable amount of protein (Bonde et al. 2011). There have been many claims that *M. koenigii* leaves can aid in weight loss. The rationale for this claim appears to be the presence of carbazole alkaloids, which are known to work against weight gain and help in regulating blood cholesterol levels in the body. Thus, in the Indian sub-continent, it is a popular belief and recommendation to consume these curry leaves. In fact, due to this property and to increase the amounts consumed, dried or fresh *M. koenigii* leaves are added in copious amounts to many dishes in restaurants. However, the credibility of this health claim is yet to be scientifically investigated.

Concluding Remarks

Research carried out on *M. koenigii* follows an ethnobotanical approach. Other parts of the *M. koenigii* plant, such as the seeds and seed oil, also possess significant medicinal benefits, yet they need to be evaluated further for their therapeutic potential, since no side effects are recorded to date. Many researchers view *M. koenigii* leaves simply as adding flavor to the food, so human clinical trials are required to substantiate health claims. This assures the safety of consuming *M. koenigii* and its potential to be promoted as an herb with multi-variate therapeutic effects. Overall, *M. koenigii* is considered a *miracle medicine* in many of the traditional medicinal systems of the Indian sub-continent, and it is hard to disagree with this notion given its variety of therapeutic compounds and their beneficial effects on human health.

ALTERNANTHERA SESSILIS LINN.

A. sessilis is an aquatic plant known by several common names in Asia as well as the rest of the world as summarized in Table 3.9. It is used as a leafy vegetable especially in Sri Lanka and some other Asian countries. A representative image of the

TABLE 3.9
Vernacular Names Used to Refer to *A. sessilis*

Language/Region	Name
Bahasa Indonesia	Daun Tolod
Bengali	Chanchi, Haicha, Sachishak
Chinese	Lian Zi Cao, Bai Hua Zi
English	Sessile joyweed
French	Brede Chevrette, Magloire
Hindi	Gudrisag, Garundi
Kannada	Honagonesoppu
Malayalam	Meenamgani, Ponnankannikkira
Malaysia	Keremak
Manipuri	Phakchet
Marathi	Kanchari
Portuguese	Bredo-D, Periquito-Sessil, Perpétua
Sanskrit	Mathsyakshi
Sinhalese	Mukunuwenna
Tamil	Ponnankanni, Citai, Koduppai
Telugu	Ponnagantikura

Source: Rao, P., *SciFed J. Herb. Med.*, 2, 1, 2018.

leaves of *A. sessilis* is shown in Figure 3.11. This herb is one of the most popular of nine leafy vegetables cultivated and sold on a commercial scale in Sri Lanka (Yapa 2015). It is easy and inexpensive to cultivate, damage during transport is minimal, and once cultivated it can be harvested each month for about two years (Senadheera et al. 2014). Sri Lankans believe that this vegetable contains high levels of vitamins, protein, and fibre and consume the plant several times per week (Senadheera and Ekanayake 2012).

A particularly interesting story relates to mis-identification of similar species elaborated by Gunasekera (2009). In the 1960s, most Sri Lankan expatriates living in Melbourne, Australia had their own patch of home grown 'mukunuwenna' – or so it was assumed. The plant attained such a rarity value and status in Melbourne that it was regularly offered to their friends and relatives in other states and territories in Australia. However, the Department of Primary Industries in Australia became aware that the Sri Lankan community had been mistakenly cultivating the wrong plant as 'mukunuwenna'. They were growing highly invasive alligator weed (*A. philoxeroides*). The similarity between alligator weed and *A. sessilis* was a serious case of mistaken identity. Alligator weed is a recognized and declared State Prohibited Weed in Australia. When this problem of mis-identification was identified by the Department of Primary Industries in Australia, they initiated a program to control the weed throughout the country. Identifying a replacement vegetable for alligator weed was an important part of this program, which also encouraged the Sri Lankan community to participate in the control effort.

FIGURE 3.11 Representative image of the leaves of *A. sessilis*, which is typically consumed in Sri Lanka.

Thus, the Department of Primary Industries in Australia decided not to promote real 'mukunuwenna' grown in Sri Lanka as a replacement plant, due to its weedy nature in other countries. An Australian native species, of the same genus commonly known as lesser joyweed (*A. denticulate*) was eventually selected as a more environmentally suitable alternative.

Origins, Morphology, and Growth

The taxonomic classification of *A. sessilis* is shown in Table 3.10. This is a perennial herb with prostrate stems, rarely ascending, often rooting at the nodes, while the leaves are obovate to broadly elliptic, occasionally linear-lanceolate, 1–15 cm long,

TABLE 3.10
Taxonomic Classification of *A. sessilis*

Kingdom	Plantae
Sub-kingdom	Viridiplantae
Division	Tracheophyta
Class	Magnoliopsida
Order	Caryophyllales
Family	Amaranthaceae
Genus	*Alternanthera*
Species	*sessilis* (L.) R. Br

0.3–3 cm wide, glabrous to sparsely villous and the petioles are 1–5 mm in length. *A. sessilis* flowers in sessile spikes with its bract and bracteoles being shiny white, 0.7–1.5 mm long, glabrous; the sepals are equal, 2.5–3 mm in length, the outer ones 1-nerved or indistinctly 3-nerved toward the base; there are 5 stamens out of which 2 are sterile. The plant flowers from December to March. Although the plant generally grows wild, it is also cultivated for food and herbal medicines as well as an ornamental plant in Sri Lanka. In Sri Lanka, the plant prefers damp, shady areas, swamps, pond margins, shallow ditches, road sides, low-lying waste places, damp pastures, and cultivated areas. However, the species can tolerate extremely dry conditions and often grows with some other aquatic species. It can also be grown for the aquarium trade in countries such as Thailand and Vietnam (though it only grows submersed for short periods), and as poultry feed.

A. sessilis is widespread throughout the tropics and subtropics. In India, it is found throughout the hotter parts, ascending to an altitude of 1,200 m in the Himalayas and even cultivated as a pot-herb (Nayak et al. 2010). The plant spreads by seed dispersal, with wind and water and by rooting at stem nodes (Singh et al. 2009). A complete protocol for micropropagation of *A. sessilis* using leaf explants was developed by Singh et al. (2009). The increased demand for *A. sessilis* plants, especially for food and medicine, has resulted in their rapid depletion from primary habitats. Thus, tissue culture techniques are being increasingly exploited for clonal multiplication and *in vitro* conservation of valuable indigenous germplasm threatened with extinction such as *A. sessilis*. Rooting behavior of stem cutting and non-availability of seeds due to over-exploitation (before flowering the leaves, *A. sessilis* are harvested for commercial purpose) is a major setback for plant propagation (Gnanaraj et al. 2011). Further, such conventional propagation processes are highly dependent on the seasons and can be achieved only during the monsoon period (Gnanaraj et al. 2011).

Traditional Medicinal Applications

A. sessilis has been traditionally used in Sri Lanka for chronic headaches, night blindness, eye infections, asthma, catarrh, urinary infections, fatigue, rabies, snake bites, and spider bites (Yapa 2015). As mentioned earlier in this chapter on herbal porridges, the leaves of *A. sessilis* are added to rice gruel and consumed. Additionally, it is also prepared in salad form in Sri Lanka, where the leaves are cut into small pieces, added with grated coconut and onions, and consumed with rice and curry. It does not have a strong taste, and therefore, blends well with any meat, fish, or vegetarian dish.

In the traditional medicinal system of India, it has been used as a blood purifier and against skin diseases and ulcers. In this medicinal system, the extract of the whole plant was used to treat infected wounds and the herb also proved styptic in colitis; its nutritive value makes the herb a potent tonic with a wide range of applications (Chopra et al. 1999; Nayak et al. 2010). A poultice of the pounded fresh plant material was used in sprains, burns, eczema, carbuncle, erysipelas, and acute conjunctivitis (Chopra et al. 1999). The extract of the herb has also been used as a cholagogue (increases bile flow in liver), abortifacient (causes abortion), and febrifuge (reduces fever) as well as to treat snakebites, dysentery, diarrhea, skin problems inflamed wounds and boils, and applied externally on acne and pimples (Chopra et al. 1999).

The traditional medicinal practitioners of the Bargarh district in India use the plant to treat dysentery with bleeding (Sen and Behera 2008).

A. sessilis is used by folk medicinal practitioners of Bangladesh for alleviation of severe pain. In Sri Lanka, it is an aquatic plant and can be commonly observed in marshy areas and wetlands (Hossain et al. 2014). Traditional medicinal practitioners of Bangladesh believe the plant possesses many medicinal properties and they have used it in many herbal formulations (Hossain et al. 2014). For example, in the Noakhali district of Bangladesh the plant is used to treat gonorrhea, low sperm count, and leucorrhea (Rahmatullah et al. 2011). In several areas of the Faridpur and Rajbari districts of Bangladesh, the plant is used by traditional medicinal practitioners for treatment of severe pain (Mukti et al. 2012). In Ghana, a decoction with some salt is taken to stop blood vomiting (Gnanaraj et al. 2011). In Nigeria, the pounded plant is used for headaches and vertigo and the leaf sap is sniffed up the nose to treat neuralgia (Gnanaraj et al. 2011). Additionally, in certain parts of Nigeria, the paste is used to draw out sharp spines or any other object from the body and is also used to cure hernias (Gnanaraj et al. 2011). In Senegal, the leafy twigs are grounded to a powder and applied on snakebites (Gnanaraj et al. 2011).

Bioactive Compounds in *A. sessilis* and Their Therapeutic Properties

Lee et al. (2014) evaluated 18 herbs including *A. sessilis* in this study on the antioxidant and starch hydrolase inhibitory activities. These two therapeutic properties were in the mid-range among the 18 herbs. The contents of lipophilic antioxidants reported in this study are shown in Table 3.11. Furthermore, anti-inflammatory constituents such as

TABLE 3.11
Lipophilic Antioxidant Compounds Present in *A. sessilis* as Reported by Lee et al. (2014)

Neoxanthin	3.69
Viola xanthin	4.58
Lutein	4.20
Zeaxanthin	0.69
Lycopene	0.49
α-carotene	0.35
β-carotene	0.47
α-tocopherol	0.28
δ-tocopherol	0.04
γ-tocopherol	0.06

Source: Lee, Y.H. et al., *J. Sci. Food Agric.*, 95, 2956–2964, 2014.

epigallocatechin, catechin, chlorogenic acid, 4-hydroxybenzoic, apigenin, vanillic acid, ferulic acid, ethyl gallate, and daidzein were reported by Othman et al. (2016) in this plant. These compounds may also contribute to the antioxidant properties of the extract from various parts of the *A. sessilis* plant. Borah et al. (2011) evaluated the antioxidant activity of different solvent extracts of *A. sessilis* whole plant using chemical antioxidant assays. Overall, the highest antioxidant activities were found in the methanolic extract, followed by the acetone extract. Quercetin was separated and identified from the hydro alcoholic extracts of *A. sessilis* stems by Vani et al. (2018). From the hydro-alcoholic extracts of stems of *A. sessilis* using GC-MS, the following major phytoconstituents, methoxy-bis (cyclopentadiene), 5,10-dihexyl- 5, 10-dihydroindolo[3,2-b]indole-2,7-dicarbaldehydeand 1,2-bis[3,4-dimethoxybenzyl]-1,2-bis(methoxymethyl)ethane, were identified respectively (Vani et al. (2017). It is hypothesized that these compounds could also demonstrate anti-asthmatic effects.

Jalalpure et al. (2008) studied the wound-healing effects in albino rats and antimicrobial properties of *A. sessilis* leaves, using different solvent extracts. In this research work, the leaves were extracted sequentially with different solvents, namely petroleum ether, chloroform, acetone, methanol, and distilled water in ascending order of polarity. All five extracts were subjected to antimicrobial screening using the cup plate and turbidimetric methods. The chloroform extract at a dose of 200 μg/mL (orally) showed significant results in rats. The presence of sterols in the chloroform extract was also confirmed by preliminary phytochemical investigation, using TLC and HPTLC methods. The wound closure properties of a stem extract of *A. sessilis* were analyzed in a study by Muniandy et al. (2018). In this study, the extract was analyzed for free radical scavenging capacity and the cell migration of two most prominent cell types on the skin, human dermal fibroblast (NHDF), keratinocytes (HaCaT), and diabetic human dermal fibroblast (HDF-D) to mimic the wound healing in diabetic patients. It was discovered in this study that the extract exhibited remarkable antioxidant, proliferative, and migratory rate in NHDF, HaCaT, and HDF-D in a dose-dependent manner, which supports its traditional medicinal application of wound-healing, and the researchers suggested it may be due to the presence of wound healing-associated phytocompounds such as hexadecanoic acid (Figure 3.12). Muniandy et al. (2018) reported the presence of several other compounds such as 2,4-dihydroxy-2,5-dimethyl-3(2H)-furan-3-one (8.92%), 2-1,2,4-trioxolane,3-phenyl- (5.99%), palmitate (5.65%), and L-glutamic acid (5.04%), although their contribution towards these therapeutic effects have not been sufficiently elucidated.

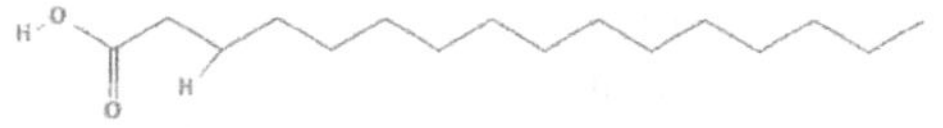

FIGURE 3.12 Hexadecanoic acid, which is hypothesized to be behind the wound-healing properties of *A. sessilis*. (From Muniandy, K. et al., *J. Immunol. Res.*, 2018.)

The suppression activity of the proinflammatory cytokines and mediators as a characteristic of anti-inflammatory action was studied by Muniandy et al. (2018) using the stem extract of *A. sessilis* in the lipopolysaccharide-stimulated RAW 264.7 macrophage cell line. The results showed that the extract significantly inhibited the production of the pro-inflammatory mediators including nitric oxide and prostaglandin-2; cytokines comprising interleukin-6, interleukin-1β, and tumor necrosis factor-α; and enzymes covering the inducible nitric oxide synthase and cyclooxygenase-2 by preventing the inhibitory protein – IκBα from being degraded, to inhibit the nuclear translocation of nuclear factor-κB subunit p65 in order to hinder the inflammatory pathway activation. Overall, the results by Muniandy et al. (2018) indicated that the stem extract of *A. sessilis* could be an effective candidate for ameliorating inflammatory-associated complications.

Hossain et al. (2014) analyzed the analgesic (non-narcotic) property of aerial parts of *A. sessilis* methanolic extract along with anti-hyperglycaemic activity. In this study, the anti-hyperglycaemic activity was measured by oral glucose tolerance tests. The analgesic activity was determined by observing any decrease in abdominal writhing in an intraperitoneally administered acetic acid-induced pain model in mice. The plant extract had successfully demonstrated its hypoglycaemic effects as well as analgesic effects in the *in vivo* model in this investigation. Gothai et al. (2018) examined the effects of three parts (aerial, leaf, stem) of the whole plant of *A. sessilis* on HT-29 colon cancer cell lines. Among these three plant part extracts, the leaf extract greatly suppressed the growth of colon cancer cells in a time- and dosage-dependent manner, followed by the aerial and stem extracts.

Concluding Remarks

Much of the presently available scientific evidence appears to support the traditional medicinal uses of *A. sessilis* for a variety of ailments such as reducing high blood glucose levels and to alleviate pain. Thus, the plant merits further scientific attention and isolation and identification of the responsible bioactive components, including demonstrating any synergistic effects with the bioactive compounds. *A. sessilis* may also be considered as an impending anti-inflammatory therapeutic agent against inflammatory-associated problems based on current scientific evidence. However, the presently available *in vitro* analysis is insufficient to be directly extrapolated to its biological activity and further animal and clinical studies are warranted. For the mass cultivation of *A. sessilis*, it is important that biotechnological approaches are used. Currently, there are few published reports on the tissue culture-based propagation of *A. sessilis*, although there are reports of micro-propagation performed on other plants from the same family. Overall, for investigative purposes, the mechanisms of action of *A. sessilis* against various disease models require further study to provide a better understanding of its abilities to effectively cure and control diseases that have a significant impact on the quality of life.

GYMNEMA SYLVESTRE R. BR.

Gymnema sylvestre R. Br. belongs to the family Asclepiadaceae. It is a perennial woody vine that grows in the tropical areas of India, Africa, and Australia and has been used for therapeutic purposes in many of the traditional medicinal systems of the tropics (Karalliedde and Gawarammana 2008). A list of vernacular names used to refer to the plant in the Indian sub-continent is shown in Table 3.12, while it is also known as Australian cowplant and Periploca of the woods (Weragoda 1980). It is interesting to note that the Hindi term 'gurmar' means 'sugar destroyer' or 'sugar blocker', and thus, the plant has been traditionally used for anti-diabetic purposes (Kurihara et al. 1969). In fact, the species name *Gymnema* appears to have been derived from the word 'gurmar' (Saneja et al. 2010). A representative image of the leaves of the plant is shown in Figure 3.13. Recently, the plant has been recognized by the natural products industry in North America and Europe and several commercial, over-the-counter herbal products are now available containing varying amounts of *G. sylvestre* (Reddy et al. 2004). *G. sylvestre* extracts have been converted at present into the forms of lozenges, mouthwash, or tea, which in turn diminishes the consumption of sweet foods and overall caloric intake (Liu et al. 2004). A few brands that contain *G. sylvestre* as an ingredient are Bioshape®, Diaxinol® (Ahmed et al. 2010), Body Slatto Tea®, Gymnema®, Gymnema Diet®, and Sugar Off® (Gopi and Vatsala 2006). *G. sylvestre* extracts formulated as mint lozenges appear to reduce the desire for high-sugar foods and the pleasant taste of candy (Stice et al. 2017). In Japan, 50 tons of *G. sylvestre* leaves are consumed annually to help with weight loss (Ueno 1993; Ogawa et al. 2004).

TABLE 3.12
Vernacular Indian Sub-continental Names That Are Used for *G. sylvestre*

Language	Name
Bengali	Meshashrunga
Gujarati gudmar	Madhunashini
Hindi	Gurmar
Kannada	Madhunashini
Konkani	Kawli
Malayalam	Chakkarakolli
Marathi	Bedakicha pala
Oriya	Lakshmi
Sanskrit	Madhunaashin
Sinhala	Masbaedda
Tamachek	Taemoerzôrt
Tamil	Sirukurinjan
Telugu	Podapatri
Urdu	Gurmar

FIGURE 3.13 Leaf structure of the *G. sylvestre* plant.

Origins, Morphology, and Growth

The taxonomic classification of *G. sylvestre* is shown in Table 3.13. *G. sylvestre* is a slow-growing herb, found ideally in tropical and subtropical humid climates and common in hills of evergreen forests (Tiwariet al. 2014). It is a climber and generally requires a solid support for its vertical growth. The seeds are sown in the months of November–December and harvested from September to February. The propagation through seed germination is difficult due to low viability of the seeds; therefore, the alternative has been root cuttings that are generally planted in the months of

TABLE 3.13
Taxonomic Classification of *G. sylvestre*

Kingdom	Plantae
Sub-kingdom	Tracheobionta
Super-division	Spermatophyta
Division	Magnoliophyta
Class	Magnoliophyta
Sub-class	Asteridae
Order	Gentianales
Family	Asclepiadaceae
Genus	*Gymnema*
Species	*sylvestre*

June and July (Reddy et al. 2004). The leaves are opposite, usually elliptic or ovate (1.25–2.0 inch × 0.5–1.25 inch); inflorescence is lateral umbel in cymes (Gurav et al. 2007). Its follicles are terete and lanceolate, up to 3 inches in height. Corolla is pale yellow in color, valvate, campanulate with single corona with 5 fleshy scales (Potawale et al. 2008). The calyx-lobes are long, ovate, obtuse, and pubescent. Carpels, unilocular, ovules locules may be present (Gurav et al. 2007; Potawale et al. 2008).

G. sylvestre is mainly grown for commercial purposes in the Deccan peninsula of western India, tropical Africa, Vietnam, Malaysia, and Japan, while in Germany and the US it is considered as a functional food (Ahmed et al. 2010). However, *G. sylvestre* natural strands are fast disappearing and are threatened with extinction due to indiscriminate collection and over-exploitation for commercial purposes, especially to meet the requirements of the pharmaceutical industry (Choudhury 1988). Commercial exploitation for production and conventional propagation is hampered due to the plant's poor seed viability, low rate of germination, and poor rooting ability of vegetative cuttings (Komalavalli and Rao 2000). Thus, alternative propagation methods are required to meet future demand in accelerating large-scale multiplication, improvement, and conservation of the plant. Komalavalli and Rao (2000) investigated an efficient and rapid propagation of *G. sylvestre* using various explants from seedling plants grown *in vitro*. The nature of the explant, seedling age, medium type, plant growth regulators, complex extracts, and antioxidants had markedly influenced the *in vitro* propagation of *Gymnema sylvestre*.

In the study by Gopi and Vatsala (2006), callus cultures were initiated from nodal segments and leaf explants of *G. sylvestre* on Murashige and Skoog (1962) medium containing basic salts and 30 g/L sucrose supplemented with different concentrations (0.10, 0.25, 0.5, 1.0, 2.0, and 5.0 mg/L) of 2,4-dichlorophenoxy acetic acid, α-naphthalene acetic acid, indole-3-acetic acid, indole-3-butyric acid, kinetin, and 6-benzyladenine. Some parts of explants were enlarged in this study and gave rise to pale yellowish calli after 2–3 weeks of incubation. cell biomass extracts were compared with extracts from leaves of naturally growing *G. sylvestre* plants. HPLC analysis of these extracts showed that the main components of the active principles, namely the gymnemic acids and gymnemagenin, were present in sufficiently large amounts in the cultured undifferentiated cells. In the study by Reddy et al. (2004), a rapid system was developed for regenerating shoots from mature nodal explants of *G. sylvestre*. Single-node stem explants were inoculated on media containing different combinations of 6-benzylaminopurine or kinetin with naphthaleneacetic acid. The maximum number of shoots (7 per explant) were observed on the medium containing 6-benzylaminopurine at 5 mg/L and naphthaleneacetic acid at 0.2 mg/L.

Traditional Medicinal Applications

Around 2–3 leaves of *G. sylvestre* are typically ground into a pulp using a mortar and pestle or a grindstone and added to rice gruel. This preparation is then consumed as a breakfast item. In the traditional medicinal system of Sri Lanka, apart from diabetes, *G. sylvestre* leaves and extracts have been used to treat eye diseases, allergies, constipation, cough, dental caries, obesity, stomach ailments, and viral infections (Ediriweera and Ratnasooriya 2009). *G. sylvestre* extract is typically consumed along with mashed fenugreek seeds and neem leaves for the treatment of diabetes, as well as with globe artichoke flowers for weight loss. In the case of hypercholesterolemia, *G. sylvestre* is recommended to be taken with turmeric, hawthorn, globe artichoke flowers, and garlic. The dosages and amounts of the herbs combined in these instances vary, depending on the severity of the condition; this knowledge typically lies with the traditional medicinal practitioner, who prepares the formula after examination of the patient and the disease condition. In other traditional medicinal systems in the Indian sub-continent, the whole plant is believed to have been used in dyspepsia, constipation, jaundice, haemorrhoids, cardiopathy, asthma, bronchitis, and leukoderma (Saneja et al. 2010). The sweet paralyzing property of the leaves was well recognized since ancient times and was used in the traditional medicinal systems of the Indian sub-continent to overcome the craving for sweets (Porchezhian and Dobriyal 2003).

In the traditional medicinal system of India, it is used in the treatment of asthma, eye complaints, inflammations, family planning, and snake bites (Selvanayagam et al. 1995; Komalavalli and Rao 2000). The drug is also used in the composition of traditional medicinal preparations in India such as *Ayaskri*, *Varunadi kasaya*, *Varunadighrtam*, *Mahakalyanakaghrtam* (Saneja et al. 2010). The ethnobotanical uses of *G. sylvestre* have been explained in detail in the review by Saneja et al. (2010). It is mentioned in this paper that there are over 400 different tribal and other ethnic groups in India. Each tribal group has their own tradition, folk language, beliefs, and knowledge about the use of natural resources as medicines. Documented usage of *G. sylvestre* is found in many of the ethnobotanical surveys among the tribes in India. Inhabitants of the Nagari Hills of the North Arcot District, Mumbai, and Gujarat appear to have the habit of chewing a few green leaves of *G. sylvestre* in the morning to keep their urine clear and to reduce glycosuria. Bourgeois classes of Mumbai and Gujarat also chew fresh leaves for the same effect. In Mumbai and Chennai, the tribe, known as Vaids, is known to recommend the leaves for the treatment of diabetes. In the same tribe, the juice obtained from the root appears to be used to treat vomiting and dysentery and in the form of paste is applied with breast milk to treat oral ulcers.

Bioactive Compounds of *G. sylvestre* and Their Therapeutic Properties

The leaves of *G. sylvestre* contain triterpene saponins belonging to the classes of oleanane and dammarane (Tiwari et al. 2014). The major constituents such as gymnemic acids and gymnemasaponins are members of the oleanane type of saponins while gymnemasides are dammarane saponins (Khramov et al. 2008). The major secondary metabolites in *G. sylvestre* include a group of nine closely related acidic glycosides, the main constituents being gymnemic acids A–D, which are found in all parts of the plant (Chakravarti and Debnath 1981). Gymnemic acids A_2 and A_3 possess both glucuronic acid and galactose in their molecular structures while glucuronic acid was found to be the only moiety in gymnemic acid A_1 (Yoshikawa et al. 1989a; Yoshikawa et al. 1989b). The derivatives of gymnemic acids are several acylated tigloyl, methylbutyryl group substituted members, derived from deacylgymnemic acid (DAGA), which is a 3-O-β-glucuronide of gymnemagenin (3β, 16β, 21β, 22α, 23, 28-hexahydroxy-olean-12-ene) (Tiwari et al. 2014). In the review by Di Fabio et al. (2014), 43 triterpenoid compounds and their analogues from *G. sylvestre* are discussed in detail for some of their anti-sweet, anti-diabetic, and anti-cancer properties, mostly verified through *in vitro* studies. However, only a few have purported beneficial effects and are selected for discussion herein.

In the study by Lee et al. (2014), *G. sylvestre* aqueous extract displayed mild antioxidant and starch hydrolase inhibitory properties. The lipophilic antioxidant contents reported in this study are shown in Table 3.14. Tannins and saponins are the chief chemical constituents present in *G. sylvestre*, which possess antioxidant as well as anti-inflammatory properties (Agarwal et al. 2000). In terms of the anti-inflammatory property, Malik et al. (2008) demonstrated that the aqueous extract of *G. sylvestre* showed predominantly significant activity, which is comparable to the standard drug phenybutazone.

In order to determine whether *G. sylvestre* contains the therapeutic potential to curb non-insulin dependent diabetes mellitus (NIDDM), Persaud et al. (1999) examined the ability of the aqueous extract of the plant to secrete insulin in rat islets of Langerhans and pancreatic β-cell lines. The study revealed the extract stimulated insulin secretion by increasing cell permeability. In the study by Ahmed et al. (2010), the methanol extract of *G. sylvestre* leaf and callus showed anti-diabetic activities through regenerating β-cells in alloxan-induced diabetic Wistar rats. Murakami et al. (1996) demonstrated that gymnemic acid moieties and its saponin constituents, which are shown in Figure 3.14, were responsible for the hypoglycaemic effects in an oral glucose tolerance test conducted in diabetic Wistar rats. Shimizu et al. (1997) also observed that extracts from *G. sylvestre* leaves had a high K^+-induced contraction of guinea pig longitudinal muscles and on glucose transport, through glucose-evoked transmural potential difference. The efficacy of *Inula racemosa* (root) and

TABLE 3.14
Lipophilic Antioxidants Present in *G. sylvestre*

Neoxanthin	14.38
Viola xanthin	5.64
Lutein	2.80
Zeaxanthin	1.24
Lycopene	0.47
α-Carotene	0.14
β-Carotene	0.61
α-Tocopherol	0.84
δ-Tocopherol	0.39
γ-Tocopherol	0.41

Source: Lee, Y.H. et al., *J. Sci. Food Agric.*, 95, 2956–2964, 2014.

Note: The data is expressed as micrograms per gram fresh weight basis ($\mu g g^{-1}$ FW) and presented as mean.

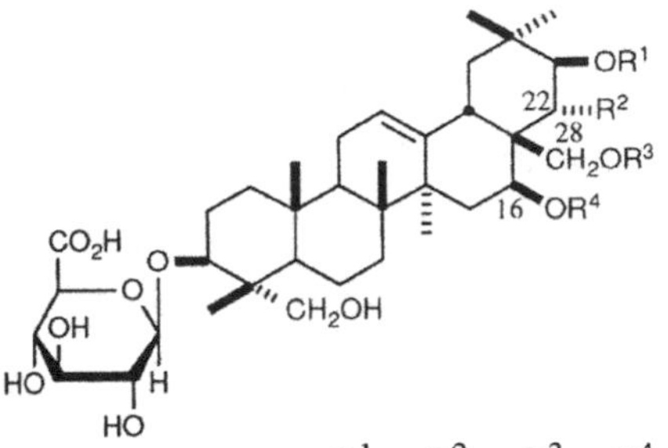

	R^1	R^2	R^3	R^4
Gymnemoside a (**1**)	Tig	OAc	H	H
Gymnemoside b (**2**)	Tig	OH	H	Ac
Gymnemic acid I (**3**)	Tig	OH	Ac	H
Gymnemic acid II (**4**)	MB	OH	Ac	H
Gymnemic acid III (**5**)	MB	OH	H	H
Gymnemic acid IV (**6**)	Tig	OH	H	H
Gymnemic acid V (**7**)	Tig	OTig	H	H
Gymnemic acid VII (**8**)	H	H	H	H
Gymnemagenin 3-*O*-glucuronide (**12**)	H	OH	H	H

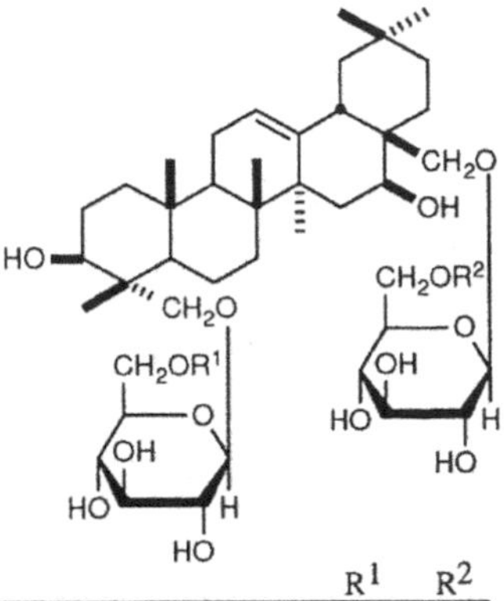

	R^1	R^2
Gymnemasaponin II (**9**)	H	H
Gymnemasaponin IV (**10**)	Glc	H
Gymnemasaponin V (**11**)	Glc	Glc

Tig : tigloyl MB : (2*S*)-methylbuthyroyl
Glc : β-D-glucopyranosyl

FIGURE 3.14 Gymnemic acid moieties and saponin constituents responsible for the hypoglycaemic effects of *G. sylvestre*. (From Murakami, N. et al., *Chem. Pharm. Bull.*, 44, 469–471, 1996.)

G. sylvestre (leaf) extracts either alone or in combination was evaluated in the amelioration of corticosteroid-induced hyperglycaemia in mice by Gholap and Kar (2003). Administration of the two plant extracts either alone or in combination decreased the serum glucose concentration in dexamethasone-induced hyperglycaemic animals. Tiwari et al. (2014) state that the primary mode of action of the gymnemic acid, which is the major phytoconstituent imparting anti-diabetic effects, is through stimulation of insulin secretion from the pancreas. It also explained therein that the compound exerts a similar effect by delaying the glucose absorption in the blood. Tiwari et al. (2014) go further into explaining that the spatial arrangements of gymnemic acids to the taste buds are similar to sugar molecules, which fill the receptors in the taste buds, thus preventing its activation by the sugar molecule in food.

In the study by Shigematsu et al. (2001), the extract of *G. sylvestre* R. Br leaves was orally administered once a day to rats fed a high-fat diet or normal-fat diet for 3 weeks to investigate its influence on lipid metabolism. It was observed that the extract did not influence body weight-gain or feed intake in both diet groups during the experimental period. It was concluded in this study that *G. sylvestre* improved serum cholesterol and triglyceride levels through influence over a wide range of lipid metabolism in rats. Thangaiyan and Ramanujam (2017) isolated a flavonoid glycoside from the flowers of *G. sylvestre* using chromatographic separation techniques and its anti-microbial activity was tested. The compound was identified as baicalein-7-*O*-glucuronide (Figure 3.15). In the study by Arunachalam et al. (2015), silver nanoparticles were bio-functionalized using the aqueous leaf extracts of

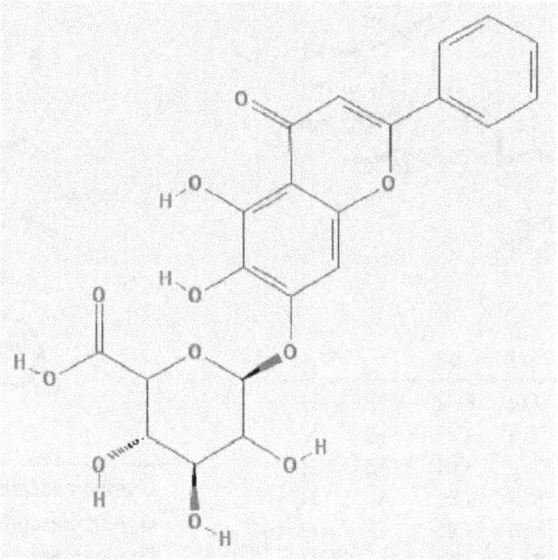

FIGURE 3.15 Chemical structure of baicalein-7-*O*-glucuronide, an antimicrobial compound from the flowers of *G. sylvestre*. (From Thangaiyan, K. and Ramanujam, S., *Papirex Ind. J. Res.*, 6, 582–584, 2017.)

G. sylvestre, and the anti-cancer properties of the bioactive compounds and the biofunctionalized silver nanoparticles were compared using the HT29 human adenoma colon cancer cell line. It was seen that the *in vitro* cytotoxic activity of the biofunctionalized green-synthesized silver nano-particles had a higher cytotoxicity of HT29 human colon adenocarcinoma cells than other cytotoxic agents and a higher cytotoxicity than the aqueous extract on its own as well.

Concluding Remarks

Diabetes is now becoming a major common disease throughout the world and many new synthetic drugs are in use and further research is ongoing. Many traditional medicinal herbs are being used in place of 'western' drugs, and it is obvious that *G. sylvestre* deserves important consideration. Despite being an important ingredient in several traditional medicinal formulations for various ailments, very few efforts have been made to verify the efficacy of *G. sylvestre* against disease conditions through scientific screening in animal models and clinical trials. The isolated bioactive compounds from *G. sylvestre* need to be further evaluated in a scientific manner using various innovative experimental models and clinical trials to understand their mechanism of action, as well as search of other active constituents, so that the multitude of therapeutic uses can be widely explored. Another void that needs to be highlighted is that although extensive research has gone into the metabolic profiling of *G. sylvestre*, there are very few reports pertaining to metabolomics and genomics. Investigations including the metabolomics and functional genomics with an emphasis on the gene identification, cloning, and their functional characterization will be an important tool in deciphering the functional role of these genes in the biochemical pathway leading to medicinal properties of the phytoconstituents of *G. sylvestre*.

AERVA LANATA (L.) JUSS. EX SCHULT

Aerva lanata (L.) Juss. ex Schult belongs to the family Amaranthaceae and is an herbaceous perennial weed growing in the warm, damp regions of the Indian sub-continent. Its taxonomic classification is shown in Table 3.15. It is commonly known as 'Pol palā' in Sinhala, 'Chaya' in Hindi, 'Bhadram' in Sanskrit, 'Mountain Knotgrass' in English, and 'Pulai' in Tamil (Krisnamoorthy 2015). As mentioned earlier in the sub-chapter on herbal porridges, the plant is commonly consumed in Sri Lanka together with rice gruel, while the herb itself is used for worm infections, urinary infections, cholera, diarrhea, bladder stones, and gonorrhea, and as a demulcent and antidote for poisoning (Tomar 2017; Rajasekaran and Gebrekidan 2018). Given its abundance and ample availability, almost all the traditional medicinal systems of the Indian sub-continent, including Sri Lanka, have some documented use in their respective pharmacopoeias for preventing and curing ailments. The difference between the medicinal systems in terms of administration of the herb is the method of preparation.

Other than being consumed in porridge form, teas prepared from *A. lanata* have become a common social beverage of food cultures in Sri Lanka (Chandrasekara and Shahidi 2018). Together with several other herbal teas and beverages, *A. lanata* has penetrated an emerging niche market along with other popular beverages such as tea, coffee, and cocoa. In addition, a rapidly growing segment of the Sri Lankan population uses this herbal beverage for slimming, weight loss, and several other cosmetic purposes (Chandrasekara and Shahidi 2018), but the scientific basis of these claims of cosmetic use is yet to be elucidated. The tea of *A. lanata* is generally prepared from leaves, stems, roots, buds, and flowers. Its commercial tea formulations may be processed in advance by means of drying, comminuting, and crushing. Figure 3.16a shows the dried plants of *A. lanata*, which are subsequently pulverized and packed for distribution and sold in supermarkets where consumers can simply add the contents of the sachets to boiling water and consume after cooling. Many commercially available *A. lanata* products are available in different forms such as whole dried plant parts, dried powder, dried particles within tea bags, as well as granulates, and solutions that can be consumed directly.

TABLE 3.15
Scientific Classification of *A. lanata*

Kingdom	Plantae
Division	Magnoliophyta
Class	Magnoliopsida
Sub-class	Caryophyllidae
Order	Caryophyllales
Family	Amaranthaceae
Genus	*Aerva* Forssk.
Species	*Aerva lanata* (L.) Juss. ex Schult.

Origins, Morphology, and Growth

A. lanata is a sub-erect, prostrate, or diffuse herb. Its stems and branches are woolly tomentose. Leaves alternate and their posterior is white and woolly, with obovate or elliptic-rounded shapes (Tomar 2017). The flowers are greenish-white and bisexual, and usually bloom in the first year of growth. The anthers are yellow, and the seeds are black and shining. Figure 3.16b shows the *A. lanata* whole plant in its natural habitat. It grows densely, covering an entire area although the plant emanates from a single stem. The plant is native to Asia and Africa. Its occurrence has been recorded in Australia as well, although this has not been recognized by the country's state herbarium (Rajanna et al. 2011). The root has a camphor-like aroma, while the dried flowers, which look like soft spikes, are sold under the commercial names 'Buikallan' and 'Boor' in the marketplace in India (Tomar 2017). *A. lanata* prefers damp sites yet is found in open forests, mountain slopes, waste and disturbed ground, deserted cultivation, and coastal scrubs. It is mostly seen at altitudes from sea level to 900 m. Its weedy nature can be more noticeably seen when it is grown in arable fields and bare patches of ground (Rajasekaran and Gebrekidan 2018).

Since *A. lanata* is considered as an important medicinal plant, it has been subjected to *in situ* conservation and population enhancement especially in India, since its natural propagation is hindered in most areas of the Indian sub-continent, due to destructive harvesting and reproductive barriers (Nandagopal et al. 2015). *A. lanata* has a narrow germplasm base and is continuously exploited for various health benefits and medicinal properties across many countries, including Sri Lanka (Rajanna et al. 2011). These authors highlight that, using conventional propagation methods, the plant takes a very long time for its development with a low rate of fruit set and poor seed germination. Moreover, this plant is naturally susceptible to fungi, bacteria, and viruses, which further reduce the yield (Nandagopal et al. 2015). In cases of degradation of land available for the cultivation of the plant (such as transformation or clearing) in the Indian sub-continent, the whole plant population in that area is wiped out.

FIGURE 3.16 (a) Dried *A. lanata* whole plants, which are generally used for making commercial products such as herbal teas, and (b) *A. lanata* whole plant in its natural moist and shady habitat.

The study by Nandagopal et al. (2015) reports the *in vitro* response of *A. lanata* explants, their reproducibility, and reliable techniques for shoot multiplication within a short period of time. Direct organogenesis of shoots from nodal segments was achieved by culturing on MS medium supplemented with different concentrations and combinations of growth regulators. It was observed that nearly 70% of *A. lanata* plantlets survived transfer to ex situ conditions in this study.

Traditional Medicinal Applications

In the traditional medicinal system of Sri Lanka, the juice obtained from *A. lanata* is primarily administered to those who have urinary infections and bladder stones (Karalliadde and Gawarmmana 2008). Today, the whole plant is generally ground into a paste using traditional kitchen equipment such as a mortar and pestle or grindstone, or modern-day kitchen blenders, and the extract is consumed in raw form or incorporated into the previously described rice gruel. The typical dosage and frequency of consumption is one teaspoon of extract twice a day for urinary infections (Karalliadde and Gawarmmana 2008). The same dose is applied in the case of bladder stones. Regardless of its traditional usage for urinary tract ailments, the porridge form is popular in Sri Lanka as a breakfast item by sports enthusiasts and health-conscious consumers, especially the young adults. The dried form can be obtained from Ayurvedic medicinal halls such as those described in Chapter 1. It is easily ground into powder and added to rice gruel. It has also been used for snake bites, especially against the venomous Common Krait (*Bungarus caeruleus*), which is a snake found mostly in the dry zone of the country (Karalliadde and Gawarmmana 2008).

'Pashanabheda' plants are a group of medicinal plants that are used in the Indian traditional medicinal system as anti-urolithiatic and diuretic drugs. *A. lanata* belongs to this group as well and sometimes the term 'Pashanabheda' is used to refer to *A. lanata* itself (Dinnimath et al. 2017). Additionally, in the same medicinal system, the extract of the herb is used as sap for eye complaints, while an infusion is given to cure diarrhea and kidney stones and the root is used as a remedy for snake bites (Krisnamoorthy 2015) – the same applications as the Sri Lankan traditional medicinal system. The Meena tribes of the Sawaimadhopur district of Rajasthan, India orally administer the juice of the roots of *A. lanata* to patients having liver congestion, jaundice, biliousness, and dyspepsia (Raja et al. 2017). These tribal people also give a decoction of the whole plant to cure pneumonia, typhoid, and other prolonged fevers (Raja et al. 2017).

Bioactive Compounds in *A. lanata* and Their Therapeutic Effects

As mentioned previously, *A. lanata* has been used in many of the traditional medicinal systems of the Indian sub-continent as a remedy for urinary problems, especially bladder stones. Thus, owing to this traditional use, many *in vivo* studies have been used to study the plant for its effects against urolithiasis. *A. lanata* from the Western Ghats of India was used in the study by Dinnimath et al. (2017) for screening anti-urolithiatic potential in male Wistar rats. The presence of quercetin and betulin in

A. lanata was believed to be responsible for a mild diuretic effect as well as the anti-urolithiatic effect by significantly reducing the size of calculi in the kidneys of the rats, and enhancing the excretion of calcium, phosphate, and oxalate while maintaining the level of magnesium, which is reported to be one of the calculi-inhibiting factors. A study by Kayalvizhi et al. (2015) investigated the nephrolithiasis property of the ethanolic extract of *A. lanata* on ethylene glycol-induced renal stones in rats. When the rats were administered with the extract at a dosage of 250 mg/kg, they had significantly lower urinary excretion and kidney retention levels of oxalate, creatinine, and calcium. The reduction of stone-forming constituents in urine and their decreased retention in the kidney contributes to the anti-nephrolithiasis property of *A. lanata*. Behera et al. (2016) isolated a hydrogenated naphthol from the flowers of *A. lanata*. This compound, (2S.3R) 3-(3-hydroxy-3methylpent-4-en-1-yl)-2,5,5,8a-tetramethyldecahydronaphthalene-2-ol was screened for anti-urolithic activity by ethylene glycol-induced urolithiasis in rats (Figure 3.17). The compound reduces the deposit of calcium oxalate crystals by increasing its solubility and restoring the normal renal architecture.

Despite the associations mentioned above between some compounds and their purported therapeutic effects, a comprehensive phytochemical analysis of *A. lanata* plant extracts has not been published to date. Thus, the study by Behera, Bag, and Ghosh (2016) appears to be the only study offering the structural elucidation of at least one of the bioactive compounds. A general phytochemical screening of *A. lanata* flower by Suganya and Subasri (2018) showed the presence of tannins, saponins, terpenoids, flavonoids, alkaloids, triterpenoids, protein, anthroquinones, polyphenol, and glycosides in methanol and aqueous extracts. Steroids were present only in the methanol extract and carbohydrates only in the aqueous extracts in this study. Adepu et al. (2013) revealed the presence of flavonoids, tannins, anthraquinones, alkaloids, phenol, proteins, amino acids, and carbohydrates. Thangavel et al. (2014) conducted a preliminary phytochemical screening and gas chromatography-mass spectrometry analysis of phytochemical constituents of *A. lanata* leaves. The presence of alkaloids, proteins, amino acids, flavonoids, tannins, phenolic compounds, saponins, quinone, terpenes, and coumarins was also reported in this study.

In the study by Hara et al. (2018), the antioxidant activities of several traditional herbs in Sri Lanka were evaluated using the DPPH free radical-scavenging assay.

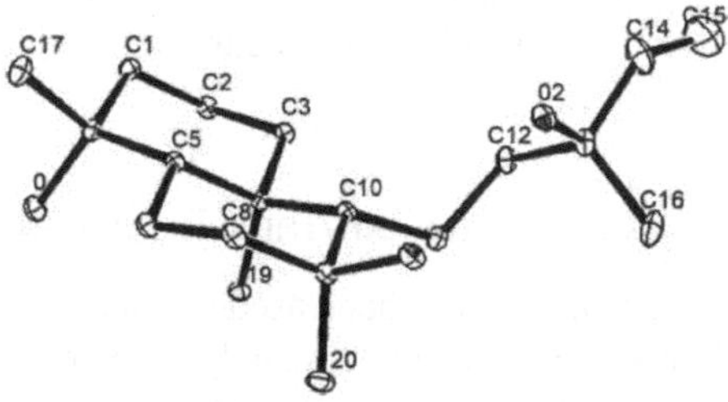

FIGURE 3.17 The chemical structure of (2S.3R) 3-(3-hydroxy-3methylpent-4-en-1-yl)-2,5,5,8a-tetramethyldecahydronaphthalene-2-ol, a hydrogenated naphthol isolated by Behra et al. (2016), which demonstrated the anti-urolithic activity of *A. lanata*.

It was shown that an *A. lanata* shoot extract obtained using 50% aqueous ethanol had mild DPPH radical scavenging effects. *A. lanata* was one of the herbs that was investigated by Jayawardena et al. (2015) for total phenolic content, total antioxidant capacity, and starch hydrolase inhibitory activity in an *in vitro* model of gastric and duodenal digestion. These three parameters were observed to be relatively stable in *A. lanata* and are retained during the digestion processes, which is an important discovery since the herb is mostly consumed orally. Typically, most bioactive compounds are expected to have a reduced bioavailability upon digestion, but it was not observed in this instance. Dinnimath and Jalalpure (2018) conducted a bioactivity guided fractionation of *A. lanata* for antioxidant and anti-urolithiatic potency using albino Wistar rats. The antioxidant assay showed that the tested fractions were as equally potent as those of standard drugs such as butylated hydroxyl toluene and Vitamin C, while the treatment with ethyl acetate and *n*-butanol fractions statistically and significantly decreased the calcium, phosphate, and blood urea nitrogen levels together with an increase in magnesium levels in rats ($p < 0.001$).

The anti-proliferative and apoptotic activity of the petroleum ether extract of *A. lanata* was investigated on Hep3B, hepatocellular cancer cells by Anusha et al. (2016). p53 mRNA expression was found to be decreased in *A. lanata*-treated cells in a dose-dependent manner, indicating the role of multiple pathways in inducing apoptosis in Hep3B cells. Krisnamoorthi and Elumalai (2018) investigated the *in vitro* anti-cancer activity of the ethyl acetate extract of *A. lanata* in the Mcf-7 human breast cancer cell line. The inhibition percentage with regards to the cytotoxicity of the extract was found to be 50% at 75 μg/mL, which was comparable to the control cyclophosphamide, which showed a cytotoxicity of 49%.

Raja et al. (2017) investigated the anti-diabetic activity of the ethanolic extract of *A. lanata* whole plant in alloxan-induced diabetic Wistar rats. The elevated blood glucose levels were reduced in the rats administered with the herb extract in this study. It was assumed that this effect was imparted due to the alkaloids present in the extract. Riya et al. (2014) explored the anti-diabetic potential of the ethanolic extract of *A. lanata* in streptozotocin-induced diabetic Wistar rats. In this study, the extract also showed a significant decrease in blood glucose levels in an oral glucose tolerance test. Akanji et al. (2018) evaluated the *in vitro* inhibitory effects of different extracts of the *A. lanata* leaves on the α-amylase and α-glucosidase inhibitory activities and chemically induced free radicals. In this study, both the aqueous ethanolic and water extracts displayed the best DPPH and ABTS radical-scavenging abilities, respectively. The activities of both α-amylase and α-glucosidase were inhibited uncompetitively by all three aqueous, ethanol, and aqueous ethanol extracts.

The increasing prevalence of multidrug resistant strains of bacteria and the recent appearance of strains with reduced susceptibility to antibiotics raise serious health concerns. Thus, it is only natural that *A. lanata* is being investigated for any anti-microbial effects. In fact, most of the traditional medicinal herbs in the Indian sub-continent that are used for various ailments have been scientifically investigated for their anti-microbial effects due to the urgent need for alternative therapies (Behera and Ghosh 2018). Murugan and Mohan (2014) investigated the anti-microbial effects of petroleum ether, benzene, ethyl acetate, methanol, and ethanol extracts of the whole plant of *Aerva lanata* against *Bacillus thuringiensis*, *Bacillus subtilis*, *Streptococcus faecalis*, *S. pyogenes*, *Staphylococcus aureus*, *Enterococcus faecalis*, *Salmonella paratyphi* A and B, *Salmonella paratyphi*, *Proteus mirabilis*, *Proteus vulgaris*, *Escherichia coli*, *Klebsiella pneumoniae*, *Pseudomonas aeruginosa*, and *P. aeruginosa* by the agar disc diffusion method. The highest degree of antibacterial activity was shown by the ethanol extract against *Klebsiella pneumonia*, while the methanol extract also showed noteworthy anti-microbial effects, specifically against *Serratia marcescens*. Devarai et al. (2017) evaluated the ethanolic and dichloromethane extracts of leaves of *A. lanata* for their anti-tubercular activity. Both extracts were investigated against a *Mycobacterium tuberculosis* H73Rv strain and the dichloromethane extract exhibited strong anti-tubercular activity with a minimum inhibitory concentration comparable to streptomycin. An overview of the main classes of antibacterial phytochemicals present in *A. lanata* and their mode of action against bacterial biofilm was presented by Rajasekaran and Gebrekidan (2018). Polyphenols in the plant were observed to have interfered with the adhesion potential, quorum sensing controlled, swarming motility, and biofilm formation of *E. coli* and *Pseudomonas aeruginosa*. Catechin and tannic acid in *A. lanata* were able to promote a significant reduction in biofilm formation by *P. aeruginosa* and were able to block biofilm formation by *E. coli* and *P. putida*.

Gujjeti and Mamidala (2014) evaluated the anti-HIV activity and cytotoxic effects of *A. lanata* root extracts. The study showed that the chloroform and methanol extracts had the highest inhibition of recombinant HIV, at 91.0% and 89.0% respectively, at a concentration of 2 mg/mL. The hexane, ethyl acetate, and acetone extractions showed highest inhibition of recombinant HIV at the same concentration of 2 mg/mL, at 86.9%, 85.2%, and 77.5%, respectively. Of note, the control drug, AZT, also showed an inhibition of 91.7% at the concentration of 2 mg/mL. This study showed the potency of *A. lanata* extracts as a potential inhibitor of HIV.

Concluding Remarks

A. lanata remains an herb of importance in modern times in Sri Lanka. Its ability to be developed into a tea as well as incorporated into rice porridge is an appealing characteristic to the urban population of the country. The increased popularity and consumption of *A. lanata* as a health drink is very encouraging as a dietary habit. From a more scientific perspective, there remains much to be discovered. The specific bioactive compounds have not been characterized to date, other than (2S.3R) 3-(3-hydroxy-3methylpent-4-en-1-yl)-2,5,5,8a-tetramethyl-decahydronaphthalene-2-ol. Despite its primary application against urinary tract ailments, given its multitude of other medicinal effects, it is obvious that many bioactive compounds exist in the plant. Existing scientific evidence demonstrating the therapeutic effects of the herb are limited to only *in vitro* and *in vivo* studies. Clinical studies are warranted to truly confirm the efficacy of the herb, especially against urinary tract-based diseases. Even though the herb has actually been used across several generations, the question always remains whether an herb that has been used for more than 3,000 years in a traditional medicinal system such as in Sri Lanka as well as in other countries of the Indian sub-continent truly requires clinical evidence as to its efficacy. It is only a matter of present-day mindset in recognizing its importance as a medication against urinary tract disorders and further investigation is warranted.

ATLANTIA CEYLANICA

There are many herbs that appear in the Sri Lankan pharmacopoeia with limited study but nevertheless remain popular among the local population. An important study by Astin (1998) revealed that patrons of complementary and alternative medicines, regardless of whether they are herbal or not, consume them because they find them to be more congruent with their own values, beliefs, and philosophies towards health and life. Despite the almost timeless battle between the traditional medicinal practices of Sri Lanka and modern science, there are some studies that show that traditional medicinal systems in general should be taken seriously (Chau 2000; Ou et al. 2003; Ko et al. 2004), even as lengthy procedures are often required to provide evidence of efficacy of traditional medicines and demonstrate and confirm the historical and folklore benefits.

A. ceylanica is considered as an under-utilized crop in Sri Lanka (Malkanthi et al. 2014). The large amount of accessible *A. ceylanica* represents a potentially massive underutilized resource, which can help to meet the increasing demand for food, energy, medicines, and industrial needs of the country (Malkanthi et al. 2014). In the case of *A. ceylanica*, it remains a fully domesticated cultivar even though much grows wild. With the development of modern agriculture practices the potential of many of these commodity resources in the country, including *A. ceylanica*, has been neglected.

Atlantia ceylanica, which is known as 'yaki nāran' in the vernacular language of Sinhala and 'pey kurundu' in Tamil (Solangaarachchi and Perera 1993), is a popular herb belonging to the citrus family used by the local population primarily for such ailments as catarrh, asthma, rheumatism, skin diseases and irritations, respiratory disorders, and to alleviate bee and wasp stings (Jayaweera 1982; Ediriweera and Ratnasooriya 2009; Yapa 2015). There is limited research available on this herb related to these health benefits. However, the herb remains a widely used domestic remedy for these disease conditions among Sri Lankans. The backyards of Sri Lanka, which have been identified as an integral part of the landscape and traditional medicinal herbal culture for centuries, remain today as one of the major areas of cultivation of *A. ceylanica*. Phytochemical studies of the bioactives in the herb are limited, but some compounds have been identified in the herbal extract. However, to date no clinical studies have been conducted.

Origins, Morphology, and Growth

A. ceylanica belongs to the family Rutaceae. The taxonomic classification of the plant is shown in Table 3.16. The plant is a dicotyledonous densely branched shrub up to 2.5 m in height, distributed in Sri Lanka and southern India (Malkanthi et al. 2014). The leaves of *A. ceylanica* are ovate, lanceolate, or elliptical (Malkanthi et al. 2014). The petiole is 3.6 mm long, thick, and glaborous, with blades mostly 4–12 cm in length (Jayaweera 1982). The inflorescence is axillary, short, cymose, or racemose (Jayaweera 1982). The leaf structure and branching of *A. ceylanica* are shown in Figure 3.18. The fruit of *A. ceylanica* is subglobose with very few vesicles and 2–4 seeds (Jayaweera 1982). It is hypothesized that this species is related

TABLE 3.16
The Taxonomic Classification of *A. ceylanica*

Kingdom	Plantae
Phylum	Tracheophyta
Class	Magnoliopsida
Order	Sapindales
Family	Rutaceae
Sub-family	Aurantioideae
Genus	*Atlantia*
Species	*ceylanica*

FIGURE 3.18 The leaf structure of *A. ceylanica*.

to *A. guillaumini*, from Indo-China, which apparently has lost its pulp-vesicles and has still larger seeds that completely fill the locules of the fruit (Reuther and Webber 1967).

Traditional Medicinal Applications

Other than the ailments mentioned in the previous chapter, *A. asiatica* leaves have been used by traditional medicinal practitioners to prepare plaster bandages (*Pattuwa*), in semi-roasted form (*Melluma*), in the preparation of warm medicinal packs and the oil extracted for massage use, (*Thaila*). According to a survey conducted by

Chulika et al. (2012), participants in the study frequently use *A. ceylanica* for diabetes mellitus – all forms of the disease condition: urinary problems, skin diseases, coughs, eye conditions, and abdominal conditions. In these instances, the leaves of the herbs are used for these ailments either in extract form or the extract incorporated into rice gruel. These are self-administered without prescription from a traditional medicinal practitioner. There appears to be no record regarding the recommended dosage of administration, although the typical amount would be one handful of leaves, ground into pulp and pressed to obtain the juice.

Bioactive Compounds in *A. ceylanica* and Their Therapeutic Effects

The presence of imperatorin and oxypeucedanin has been reported in the leaf extract of *A. ceylanica* (Bacher et al. 1999) (Figure 3.19). Imperatorin exerts a strong hepatoprotective activity (EC_{50} = 36.6 μM) against tacrine-induced cytotoxicity in HepG2 cells (Oh et al. 2002). Imperatorin was also found to inhibit HIV-1 virus replication in T- and HeLa cells by targeting the critical transcription factor Spl, and to stop infected cells at the G(1) phase of the cell cycle (Sancho et al. 2004). Reisch et al. (2004) reported that the compound oxypeucedanine weakly inhibited the growth of *Staphyloccocus aureus*. In an *in vitro* study by Tada et al. (2002), oxypeucedanine exhibited moderate to strong inhibition of cytokine release at a concentration of 10 μg/mL for Interleukin (IL)-2 and tumor necrosis factor-α, but inactive against IL-4. In RAW 264.7 macrophages stimulated with lipopolysaccharide (1 μg/mL), oxypeucedanine inhibited (IC_{50} = 16.8 μg/mL) nitric oxide (NO) production, although it exerted no activity on iNOS (Wang et al. 2000).

Bowen and Patel (1987) isolated several compounds including acridone alkaloids from the leaves of *A. ceylanica*, including 2,4,5-trimethoxybenzaldehyde, 1,5-dihydroxy-3-methoxy-10-methyl-9(*10H*)-acridone, 11-hydroxynoracronycine and pyranoflavone

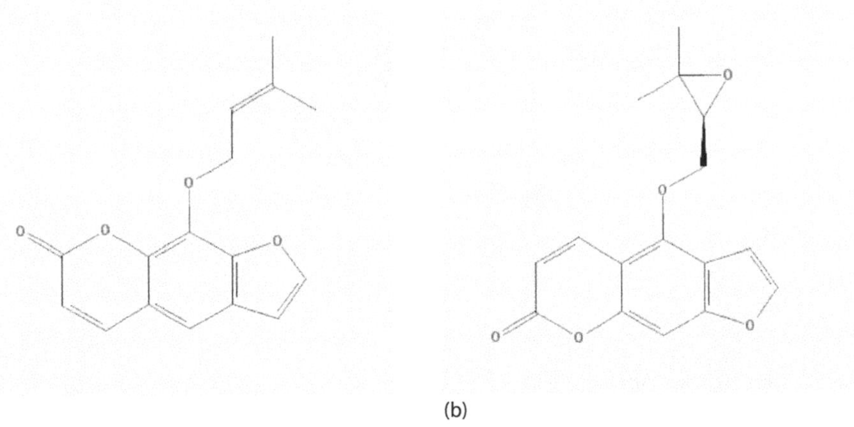

FIGURE 3.19 (a) Imperatorin and (b) oxypeucedanin, which is found in the leaf extract of *A. ceylanica*. (From Curini, M. et al., *Curr. Med. Chem.*, 13, 199–222, 2006.)

carpochromene (Figure 3.20). Fraser and Lewis (1973) isolated 3,12-dihydro-6,11-dihydroxy-3,3,12-trimethylpyrano[2,3-*c*]acridin-7-one and its 5-(3-methylbut-2-enyl) derivative from the leaf extract of *A. ceylanica*. However, the bioactivity of these compounds has not been systematically studied to date.

Fernando and Soysa (2015) conducted a study on H_2O_2 radical scavenging activity of five herbal extracts that were prepared separately as decoctions according to the proportions set out by Ayurvedic practitioners. It was reported therein that out of the 5 herbs of *A. ceylanica*, *Eriocaulon quinquangulare*, *Semecarpus parvifolia*, *Mollugo cerviana*, and *Camellia sinensis*, *A. ceylanica* ranked fourth in terms of its H_2O_2 scavenging ability. Furthermore, Fernando and Soysa (2014) conducted an *in vitro* study of the phytochemical constituents in the lyophilized powder of the water extract of *A. ceylanica* leaves for antioxidant and hepatoprotective activity.

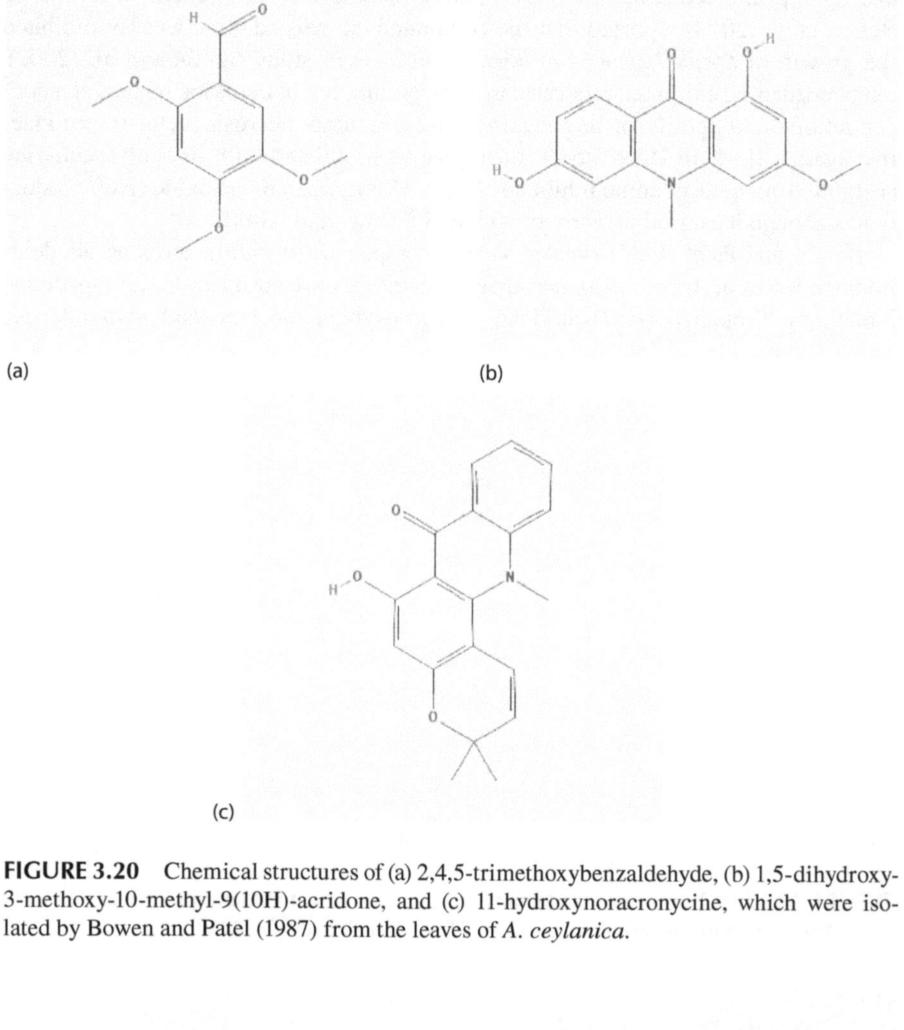

(a) (b)

(c)

FIGURE 3.20 Chemical structures of (a) 2,4,5-trimethoxybenzaldehyde, (b) 1,5-dihydroxy-3-methoxy-10-methyl-9(10H)-acridone, and (c) 11-hydroxynoracronycine, which were isolated by Bowen and Patel (1987) from the leaves of *A. ceylanica*.

In this experiment, hepatotoxicity was first induced with ethanol on porcine liver slices to study the hepatoprotective activity of the powder. Overall, a reduction in the lipid peroxidation was observed in liver slices treated with the leaf extract. Napagoda et al. (2016) evaluated the photoprotective activity of aqueous extracts (1 mg/ml) of 11 medicinal plants in Sri Lanka, including *A. ceylanica*, that have been widely employed in traditional medicine as treatment options for various skin diseases and to improve skin complexion. Among the extracts, *Atlantia ceylanica* (along with *Hibiscus furcatus*, *Leucas zeylanica*, *Mollugo cerviana*, *Olax zeylanica*, and *Ophiorrhiza mungos*) possesses sun protection-factor values of more than 25, which were even higher than two commercial photoprotective creams used in the study as reference compounds.

Munasinghe et al. (2015) investigated the antibacterial activity of the aqueous extract of leaves of *A. ceylanica* against *Pseudomonas aeruginosa*, *Streptococcus pneumonia*, *E. coli*, *S. aureus*, and methicillin-resistant *S. aureus* (MRSA), but none of these strains was inhibited. Similarly, Bandara et al. (1990) evaluated the antimicrobial and insecticidal activity of 10 plant species and concluded that the steam distillate of *A. ceylanica* only demonstrated weak pesticidal activity.

Only a few studies looked at the non-nutritional, beneficial aspects of *A. ceylanica*. In one study by Ranaweera (1996), *A. ceylanica* was one of the 53 plants tested for larvicidal activity against *Culex quinquefasciatus*, *Aedes aegypti*, *Aedes albopictus*, *Anopheles culicifacies*, and *Anopheles tessellatus* larvae. However, the plant extract exhibited very weak larvicidal activity as compared with the rest of the plant species used for the study. *A. ceylanica* was also one of the plants included in the larval food plant study by Jayasinghe et al. (2014). The impact of introducing new plants on the dynamics of abundance and distribution of butterflies is an important area of research in Sri Lanka given that the country is inhabited by 245 butterfly species including 26 endemic species. Most likely owing to its citrus aroma, *A. ceylanica* appears to be a larval food plant of interest to some of the butterfly species of Sri Lanka, and thus, the plant could be considered in conservation management of the butterfly fauna in the country. *A. ceylanica* is one of nine herbs used in the traditional herbal formulation for de-worming calves and elephants in captivity (Piyadasa 1994).

Concluding Remarks

A. ceylanica remains a very important herb in the Sri Lankan traditional medicinal pharmacopoeia, despite a limited number of scientific studies. It is sometimes grown in the backyards of Sri Lankan households, simply for the citrus smell emanating from the leaves, which adds a pungent fragrance to the gardens. There are many gaps and voids when it comes to elucidating the beneficial effects and bioactive compounds present in *A. ceylanica*. In fact, it is considered an under-utilized herb, and systematic studies would better justify its use as a crop worth cultivation. Alternative medications such as *A. ceylanica* hold much promise, due to their history and acceptance in modern times, but since their true effects are not scientifically proven, it is evident that further investigations should be performed to assess the clinical benefits of such herbs.

SESBANIA GRANDIFLORA (L.) PERS

Sesbania grandiflora is a commonly used traditional medicinal herb in Sri Lanka. It is known as 'kathurumurunga' in Sinhala and 'agathi keerai' in Tamil in the local vernacular languages. In India, in Bengali, it is known as 'agate', while it is known as 'swamp pea' in English. Various literature has indicated that there are around 60 global species belonging to the genus *Sesbania*, which are commonly found in Africa, Australia, and Asia (Sathasivam and Lakshmi 2017). However, *S. grandiflora* is commonly found in the tropical regions (Wagh et al. 2009). In Sri Lanka, usually the leaves of *S. grandiflora* are found in the marketplace. Representative images of the leaves that are bunched together and sold in various stores and supermarkets, as well as by vendors, are shown in Figure 3.21. The bark and flowers of the plant may also be found on offer among some of the street vendors in Sri Lanka, although rarely since most people tend to pick these parts fresh directly from the trees.

The tender leaves, green fruit, and flowers are eaten in Sri Lanka as a vegetable or mixed into curries or salad. A particularly favorite dish uses the flowers that have not yet bloomed and tend to be softer and less leathery, which are dipped in batter and fried in oil. The resulting dish is crispy and crunchy, almost like a snack. The flowers might also be bitter and astringent, but it is possible to reduce the bitterness by removing the stamen.

There are many miscellaneous uses of *S. grandiflora* in modern times other than for its traditional medicinal properties. For instance, it is reported as a potential pulpwood (Wei et al. 2018). In the study by Wei et al., the sodium hydroxide effect on the

FIGURE 3.21 *S. grandiflora* leaves in a bunch, which could be bought in supermarkets and from street vendors in Sri Lanka.

paper made from pulp derived from *S. grandiflora* was investigated. Sodium hydroxide is a common chemical used as part of full-stage chemical bleaching of industrial pulp. Wei et al. (2018) bleached the pulp obtained from *S. grandiflora* with sodium hydroxide at 3%, 6%, and 9% based on pulp weight, respectively. It was observed overall that sodium hydroxide has the potential to improve the optical properties of *S. grandiflora* pulp and its mechanical properties. Additionally, since *S. grandiflora* is a fast-growing tree species with a high biomass, it has been studied for its application for phyto-remediation purposes, especially against accumulation of lead (Malar et al. 2014). The study by Malar et al. (2014) aimed at evaluating phytotoxicity of various concentrations of lead nitrate (0, 100, 200, 400, 600, 800, and 1000 mg/L) in *S. grandiflora*. Outcomes of the study indicated that higher concentrations of lead enhanced the oxidative damage by over-production of free radicals in *S. grandiflora* that had a potential tolerance mechanism to lead. Also, retention of high levels of lead in the root of the plant indicated that *S. grandiflora* has the potential for phyto-extraction of heavy metals by rhizo-filtration.

Soil salinity is a widespread problem, restricting plant growth and biomass production especially in arid, semiarid, and tropical areas. The development of salt-tolerant crops or desalination of soil by leaching excessive salts, though successful, is not economical for sustainable agriculture. In this respect, biological processes such as mycorrhizal application to alleviate salt-stress and use of moderately salt-tolerant tree species are better options. *S. grandiflora* is a moderately salt-tolerant legume tree and is commonly used for overcoming this salt-stress problem (Giri and Mukerji 2004). *S. grandiflora* trees have been used as fodder throughout Indonesia especially during dry season feeding of cattle and goats (Giri and Mukerji 2004; Malar et al. 2014). Cutting management is critical to the productivity of *S. grandiflora* since it cannot survive repeated cutting. In Indonesia, farmers have devised a system where only the side branches of the tree are cut for fodder, leaving the main growing stem untouched (Wei et al. 2018).

Origins, Morphology, and Growth

S. grandiflora is believed to be a native of Malaysia although it is more abundantly found and consumed in the Indian sub-continent, especially the Indian regions of Punjab, Delhi, Bengal, Assam, and the Andaman Islands, as well as Sri Lanka. The taxonomic classification of the herb is shown in Table 3.17. It is a small, erect, quick-growing, short-lived, soft-wooded tree that is sparsely branched (Wagh et al. 2009). Its bole is straight and cylindrical, while the wood is white and soft. The tree is 5 to 12 m in height (Wagh et al. 2009). The leaves are 20 to 30 cm long and pinnate, having 20 to 40 pairs of leaflets that are 2.5 to 3.5 cm long (Wagh et al. 2009). The flowers are white and 7 to 9 cm in length; the pods are linear, 20 to 60 cm in length, 7 to 8 m wide, pendulous, and somewhat curved, containing many seeds (Wagh et al. 2009).

S. grandiflora tree has an early rapid growth and erect habit that usually tries to access sunlight by topping over neighboring plants. The roots of *S. grandiflora* are heavily nodulated. During waterlogging and floods, the tree develops floating roots and protective spongy tissues, although it is intolerant of high winds that can

TABLE 3.17
Taxonomic Classification of *S. grandiflora*

Kingdom	Plantae
Division	Tracheophyta
Sub-division	Spermatophytina
Class	Magnoliopsida
Order	Fabales
Family	Fabaceae
Genus	*Sesbania*
Species	*Sesbania grandiflora* (L.) Pers

break its stems and branches. The lifespan of *S. grandiflora* is about 20 years (Wagh et al. 2009). It can be propagated by seeds or plant cuttings.

The growth and cultivation of the plant is an aspect that is being comprehensively studied, especially in India, given that many environmental and soil conditions affect its development, and also since the plant is considered as a source of income for many tribes of the South Asian region (Dhanapackiam and Ilyas 2010). As highlighted in the previous section, there are many non-medicinal uses of *S. grandiflora* as well, which has triggered the investigation of more effective ways of cultivating the plant through biotechnological means. Reports that have successfully demonstrated such cultivation methods are scarce at the moment, although this is an area for further research and development concerning the plant.

Traditional Medicinal Applications

All parts of the plant have been recorded to be used for various ailments in the traditional medicinal systems of South Asia. *S. grandiflora* has been long used in the traditional medicinal system of Sri Lanka for the treatment of diarrhea, snake bite, malaria, smallpox, fever, scabies, ulcer, and stomach disorders (Bhalke et al. 2010). As mentioned earlier in the sub-chapter on herbal porridges, the leaf extract has been incorporated into rice gruel in addition to being consumed in isolation, for various ailments such as fever, cough, asthma, and catarrh, improving memory power, cracked lips, joint swellings, and night blindness. The porridge form may not be all that popular in modern times, though, as compared with *C. asiatica* for instance, since it possesses an astringent and bitter flavor that may not be palatable. Overall, in all traditional medicinal systems of South Asia, *S. grandiflora* leaves are believed to possess cooling, tonic, anthelmintic, and antipyretic properties (Bhalke et al. 2010). The fruits of *S. grandiflora* have been used as a laxative and a stimulant. It has also been used in the treatment of anaemia, bronchitis, fever, pain, thirst, and tumors in the Ayurvedic medicinal system of India (Bhalke et al. 2010). The root has been used in this medicinal system for inflammation and as a remedy for night blindness. Additionally, the plant serves as a leafy vegetable species and an important source of nutrients for the rural tribes of India residing in areas of Jharkhand (Sinha 2018).

The *S. grandiflora* flower has been used in the traditional medicinal systems of both India and Nepal for the treatment of diabetes mellitus (Kumar et al. 2015). However, a clear record indicating its usage for this ailment in Sri Lanka has not surfaced to date. Due to the awareness of the scientific evidence, in modern times, the leaves of the plant are consumed by many in Sri Lanka for the prevention and control of diabetes mellitus, both types I and II. Typically, the leaves are consumed in salad form, where one handful of leaves is cut and added with grated coconut and eaten with the staple diet of rice and curry. However, due to its excessive cooling effect mentioned in the traditional medicinal applications, the salad is consumed only once per day, even though the origins of this dosage have not yet been traced.

Bioactive Compounds in *S. grandiflora* and Their Therapeutic Effects

It is reported that various parts of *S. grandiflora* are rich with various phytoconstituents such as saponins, flavonoids, phenolics, alkaloids, tannins, carbohydrates, proteins, and glycosides, which are pharmacologically active (Malik and Nayak 2011; Nandi et al. 2016). Galactomannans, oleanolic acid, and β-sitosterol have also been reported in the plant (Panigrahi, Panda and Patra 2016). A preliminary phytochemical screening conducted by Kundu et al. (2018) showed that the methanolic extract of the leaves contains many flavonoids and accounts for the therapeutic properties exhibited by the leaves against many ailments. Three isoflavanoids, isovestitol, medicarpin, and sativan, along with betulinic acid, were isolated from the root of *S. grandiflora* by Hasan et al. (2012). The chemical structures of these compounds are shown in Figure 3.22. The anti-tuberculosis effect of these compounds was also

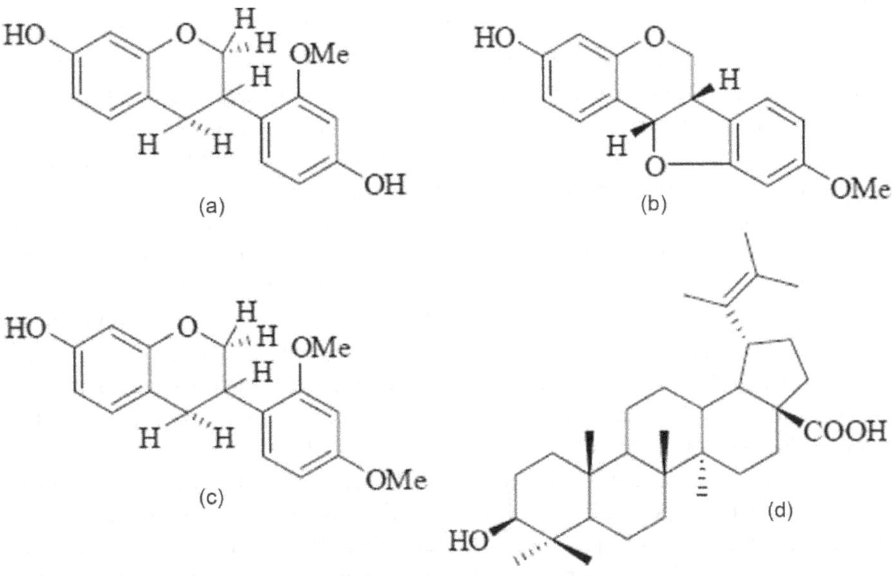

FIGURE 3.22 (a) Isovestitol, (b) medicarpin, (c) sativan, and (d) betulinic acid isolated from the bark of *S. grandiflora*. (From Hasan, N. et al., *Pharmaceuticals*, 5, 882–889, 2012.)

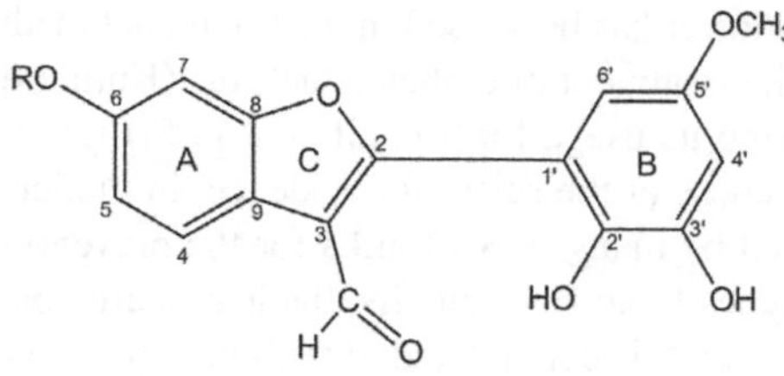

(1) R = OCH_3
(2) R = H

FIGURE 3.23 Sesbagrandiflorain A and B isolated from the ethyl acetate fraction of the stem bark of *S. grandiflora* by Noviany et al. (2018).

verified and confirmed *in vitro* in this study. Noviany et al. (2018) successfully isolated two new 2-arylbenzofurans – sesbagrandiflorain A and B – from the ethyl acetate fraction of the stem bark of *S. grandiflora*. The chemical structures of these compounds are shown in Figure 3.23. These compounds are reportedly present in the genus *Sesbania* and other members of the family Fabaceae, but the bioactivity of both these isolated compounds is yet to be elucidated.

Kumar et al. (2015) evaluated the anti-diabetic activity of the ethanolic extract of *S. grandiflora* flower in alloxan-induced diabetic rats. At dosages of 250 and 500 mg/kg significant ($p < 0.01$) anti-diabetic activity was observed compared with the diabetic control, glibenclamide. Both these dosages showed significant ($p < 0.01$) reduction of serum total cholesterol and triglyceride levels in this study. Panigrahi, Panda, and Patra (2016) have also observed the anti-diabetic activity of the methanolic extract of *S. grandiflora* leaves in type 2 diabetic rats induced by low-dose streptozotocin and a high-fat diet. The dosages of 200 and 400 mg/kg given to the experimental animals resulted in a significant reduction ($p < 0.05$) of raised blood glucose levels and restored other parameters to normal levels. Kothari et al. (2017) investigated the α-amylase inhibitory effect of *S. grandiflora* leaf extracts, where a maximum inhibition of 81% was observed at 1000 μg/mL of extract, which was compared with that of standard acarbose showing 93% inhibition. Inhibition of α-amylase is important in the realms of controlling post-prandial hyperglycaemia, where the enzyme contributes to the release of glucose to circulating blood. Guillasper et al. (2015) investigated the hypoglycaemic activity of the aqueous extract of *S. grandiflora* flowers among non-human primates (i.e. the Philippine Macaque/*Macaca fascicularis Philippinensis*), which is said to have a close genetic relationship to humans. It was observed that the extract and insulin have similar effects on the blood glucose levels of the tested animals, which can reduce almost 90%–95% of the blood glucose levels in the monkeys.

In the study by Anantaworasakul et al. (2017), the ethyl acetate fraction of *S. grandiflora* bark extract showed high antioxidant activity along with free radical scavenging and reducing mechanisms. The antioxidant effect of the *S. grandiflora* leaves was associated with its ability to prevent damage from cigarette smoke upon the lungs in the *in vivo* study by Ramesh et al. (2007). In a similar study by Ramesh et al. (2010), the protective mechanism of the aqueous extract of *S. grandiflora* leaves against cigarette smoke-induced oxidative damage in liver and kidney of adult male Wistar-Kyoto rats was demonstrated. The elevated hepatic, renal, and lipid peroxidation markers were decreased while the diminished antioxidant levels were restored. The ethanol and acetone extracts of *S. grandiflora* white variety (flower and leaf) were examined for radical scavenging capacities and antioxidant activities by Siddharaju et al. (2014). Though both extracts exhibited dose-dependent reducing power activity, the acetone extract of all the plant samples was found to have a higher hydrogen donating ability. The potent antioxidant activities of the aqueous extract of the leaves of *S. grandiflora* were reported in the study by Waisundara and Watawana (2014) as well.

Pajaniradje et al. (2014) studied the anti-cancer properties of the leaves of *S. grandiflora*. In this study, five different solvent fractions from the leaves of *S. grandiflora* were tested on cancer cell lines such as MCF-7, HepG2, Hep-2, HCT-15, and A549. The methanolic fraction of *S. grandiflora* was found to exert potent anti-proliferative effects especially in the human lung cancer cell line, A549. Caspase-3 was activated in the methanolic fraction treated A549 cells, thereby leading to cell death by apoptosis. Evidence for the anti-cancer efficacy of a protein fraction isolated from the flower of *S. grandiflora* was reported by Laladhas et al. (2010). The fraction was evaluated in two murine ascites tumor cell lines and human cancer cell lines of different origin for its anti-cancer effect. The fraction had inhibited cell proliferation and induced apoptosis as demonstrated by DNA fragmentation and externalization of phosphatidyl serine in Daltons lymphoma ascites (DLA) and colon cancer cells (SW-480).

China et al. (2012) tested the antimicrobial effects of the polyphenolic extracts of the edible flower of *S. grandiflora* against some common pathogenic bacteria and growth promoting property against the probiotic organism *Lactobacillus acidophilus*. The *in vitro* study suggested that the extract has an inhibitory effect against *S. aureus*, *Shigella* spp., *Salmonella Typhi*, *Escherichia coli*, and *Vibrio cholerae*. *S. grandiflora* extract induced a significant biomass increase of *Lactobacillus acidophilus* grown in liquid culture media in this study. The research work by Anantaworasakul et al. (2011) demonstrated that the crude ethanolic extracts obtained from different parts of *S. grandiflora* exhibited various antibacterial activities. The ethyl acetate or butanol extract of the stem bark possessed the most pronounced antibacterial activity. The kinetic study of bactericidal activities revealed that the butanol-fractionated extract of the stem bark was effective against Gram-negative bacteria (Anantaworasakul et al. 2011). Additionally, in the study by Anantaworasakul et al. (2017a), the ethyl acetate fraction presented a high potential of inhibiting bacterial growth with the minimum bactericidal concentration being less than 1 mg/mL. In the study by Anantaworasakul et al. (2017b), healthy silkworms were exposed to three fractionated extracts (3.1–400 mg/mL) of *S. grandiflora* bark from hexane, chloroform, and ethyl acetate. The ethyl acetate fraction showed the highest activity with a minimum inhibitory concentration against MRSA and vancomycin-resistant enterococci (VRE) of 1.6 and 0.4 mg/mL, respectively.

S. grandiflora leaves have been known throughout the generations for their anthelmintic property. This beneficial effect was investigated and confirmed in the study by Karumari et al. (2014), where steroids and triterpenoids were observed to be present in the extract and were associated with this property. Bhalke et al. (2010) investigated the anti-ulcer activity of the ethanolic extract of the leaves of *S. grandiflora*.

In this study, acute gastric ulceration was produced in rats by the oral administration of various noxious chemicals including aspirin or ethanol or indomethacin. The extract was administered at a dosage of 400 mg/kg orally. A significant reduction in the ulcer index was observed in this instance. In a similar study, Sertié et al. (2001) observed that the ethanolic extract of the bark of *S. grandiflora* prevented acute gastric injury in rats at a dosage of 36.75 mg/kg. Sathasivam and Lakshmi (2017) investigated the cytotoxic effects of the leaves of *S. grandiflora*. It was observed that at a concentration of 10 mg/ml, 100% mortality was seen in the brine shrimp in 24 hours. Semwal et al. (2018) evaluated the neuroprotective effect of the ethanolic extract of *S. grandiflora* seeds in mice. The ethanolic extract of *S. grandiflora* had improved the cognitive dysfunction in celecoxib-treated mice through the modification in the cholinergic system or by the blockage of oxidative stress and inhibition of acetyl-choline esterase enzyme.

Concluding Remarks

Unlike many of the other traditional medicinal herbs of Sri Lanka, it is apparent that *S. grandiflora* has many uses other than for its application for the prevention and containment of disease conditions. Phyto-extracting heavy metals by rhizofiltration is a valuable benefit of *S. grandiflora* and serves as a justification for its mass-cultivation, not only as a food source but also for benefiting the environment. In terms of the therapeutic effects of the plant, as seen from the evidence presented herein, there is limited human clinical investigation. There is also an opportunity to identify and characterize the bioactive compounds of *S. grandiflora*, especially since some of the novel compounds isolated to date have not been investigated for their therapeutic effects. In Sri Lanka, *S. grandiflora* remains an important part of the daily diet of the local population as a leafy vegetable. It may not always be consumed for purposes of preventing diseases, but simply for the maintenance of health and wellness. The salad of *S. grandiflora* leaves and dishes made from the flowers remains a signature culinary cuisine of the country, while the tree may be grown in backyards for mere ornamental purposes.

VERNONIA CINEREA (L.) LESS

V. cinerea is commonly known as 'monarakudumbiya' in Sinhala, 'little ironweed' in English, 'joanbeer' and 'kukshim' in Bengali, 'puvamkurunnel' in Malayalam, and 'sahadevi' in Sanskrit and Hindi (Abirami and Rajendran 2012). This is yet again another herb that is considered as a weed and is mostly consumed in the Sri Lankan Ayurvedic medicinal system by adding to rice gruel. The juice of the whole plant may also be used for various ailments; however, for reasons described previously, consumption of the porridge form of the herb is more popular among the local population for maintenance of health and wellness. *V. cinerea* grows wild in many areas of the Sri Lankan outdoors. It is interesting to note that this is a popular herb for domesticated cats as a remedy for GI problems, due to consumption of infected meat. Thus, among all the herbs growing in Sri Lanka the harvesting of *V. cinerea* needs care to be sure that parts of the plant have not already been consumed by cats.

Origins, Morphology, and Growth

The taxonomic classification of *V. cinerea* is shown in Table 3.18. Akobundu and Agyakwa (1998) provided some descriptive morphological characteristics of *V. cinerea*. It is a terrestrial, annual, and erect herb. The stem is erect and thin with mainly vertical branching, usually ribbed, and contains short fine hairs. The leaves are simple, alternate, and variable from 2–8 cm long and 2–3 cm wide. The lower leaves are ovate with entire or sub-entire margins, while the upper leaves are small, linear, and irregularly toothed. The petioles are short and winged. The inflorescence is a flat-topped panicle with many round, purple, or pinkish-blue florets about 3 mm across, clustered together on slender stalks and usually with numerous involucres. The fruit is an achene about 1.5 mm long with white pappus. It propagates by seeds, and flowering and fruiting can be seen throughout the year.

V. cinerea is found throughout the Indian sub-continent, Africa, and Australia (Yapa 2015) and grows in the middle and southern areas of Taiwan (Kuo et al. 2003). In Sri Lanka, it can be seen in dry grassy areas (Figure 3.24a), along roadsides (Figure 3.24b), in open waste places, and in agricultural plantations. It can also grow in wet and dry soil and in rock crevices, and at altitudes up to 800 m. During the rainy season in Sri Lanka, the growth of the plant is accelerated and can grow even up to a height of 1 m; however, by the end of the rainy season, the plant shows a reduction in size, forming small diffuse habit.

Traditional Medicinal Applications

In Sri Lankan Ayurvedic medicine, the juice of the whole plant of *V. cinerea* is used for the treatment of conjunctivitis, rheumatism, appendicitis, encephalitis, skin diseases, prolonged fever, cough, and cold (Yapa 2015). In the traditional medicinal system of India, *V. cinerea* has been traditionally used for febrifuge, diaphoretic, diuretic, and antispasmodic and anthelmintic effects (Khare 2007). The anti-malarial activity of *V. cinerea* has triggered its use against this ailment in Burkina

TABLE 3.18
Taxonomic Classification of *V. cinerea*

Kingdom	Plantae
Sub-kingdom	Tracheobionta
Division	Magnoliophyta
Sub-division	Magnoliopsida
Sub-class	Asteridae
Order	Asterales
Family	Compositae
Genus	*Vernonia*
Species	*cenerea*

FIGURE 3.24 *V. cinerea* as seen (a) at dry grassy sites and (b) along roadsides in Kandy, Sri Lanka.

Faso as well (Soma et al. 2017). In Cambodia, *V. cinerea* is called 'Kbal Ruy' by the local population and is widely used as a traditional medicine for the treatment of fever and colic (Perry 1980; Dy 2000). An ethnobotanical survey in different regions of Cambodia indicated that this plant was found to be used as a febrifuge against malaria (Hout 2006). The plant has been documented and recommended in Thai traditional medicine for smoking cessation and relief of asthma, cough, fever, malaria, urinary calculi, and arthritis (Leelarungrayub et al. 2010).

Bioactive Compounds of *V. cinerea* and Their Therapeutic Effects

Bioassay-directed fractionation of the ethanolic extract of the stems of *V. cinerea* conducted by Kuo et al. (2003) resulted in the isolation of two novel sesquiterpene lactones: vernolide-A and -B (Figure 3.25). Biological evaluation in this study showed that vernolide-A demonstrated potent cytotoxicity against human KB, DLD-1, NCI-661, and Hela tumor cell lines while vernolide-B had marginal cytoxicity in all these cell lines. Chea et al. (2006) isolated two new sesquiterpene lactones, vernolides C and D, and six known ones were isolated from the dichloromethane fraction of an aqueous extract from *Vernonia cinerea* (Figure 3.26). Preliminary phytochemical screening of *Vernonia cinerea* (Family: Asteraceae) conducted by Haque et al. (2012) showed the presence of steroids, glycosides, triterpenoids, and esters in the methanolic extract of the stem bark and leaves of the *V. cinerea*. NMR data also confirmed the presence of lupeol, 12-oleanen-3-ol-3β-acetate, stigmasterol, and β-sitosterol in the n-hexane extract of the plant

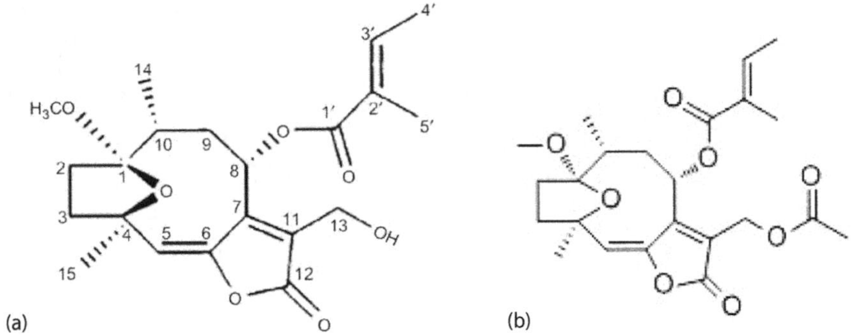

FIGURE 3.25 Chemical structures of (a) vernolide-A and (b) vernolide-B isolated by Kuo et al. (2003) from the ethanolic extract of the stem of *V. cinerea*, which showed anti-cancer properties.

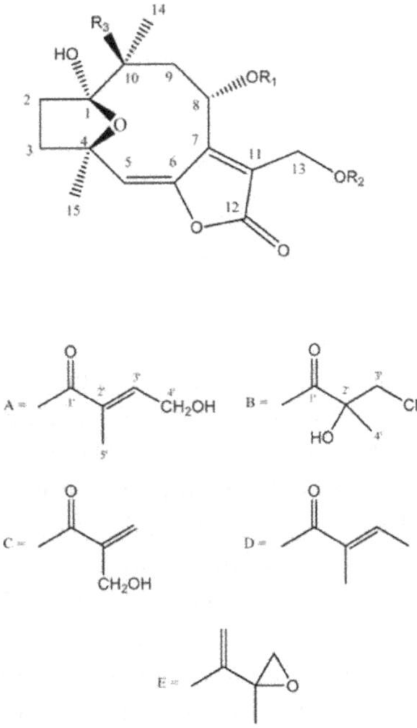

	R_1	R_2	R_3
1	D	Ac	H
2	E	Ac	H
3	D	H	H
4	B	Ac	H
5	H	Ac	H
6	H	Ac	OH
7	C	Ac	H
8	A	Ac	H

FIGURE 3.26 Sesquiterpene lactones from *V. cinerea* with anti-malarial activities isolated by Chea et al. (From Chea, A. et al., *Chem. Pharm. Bull.*, 54, 1437–1439, 2006.)

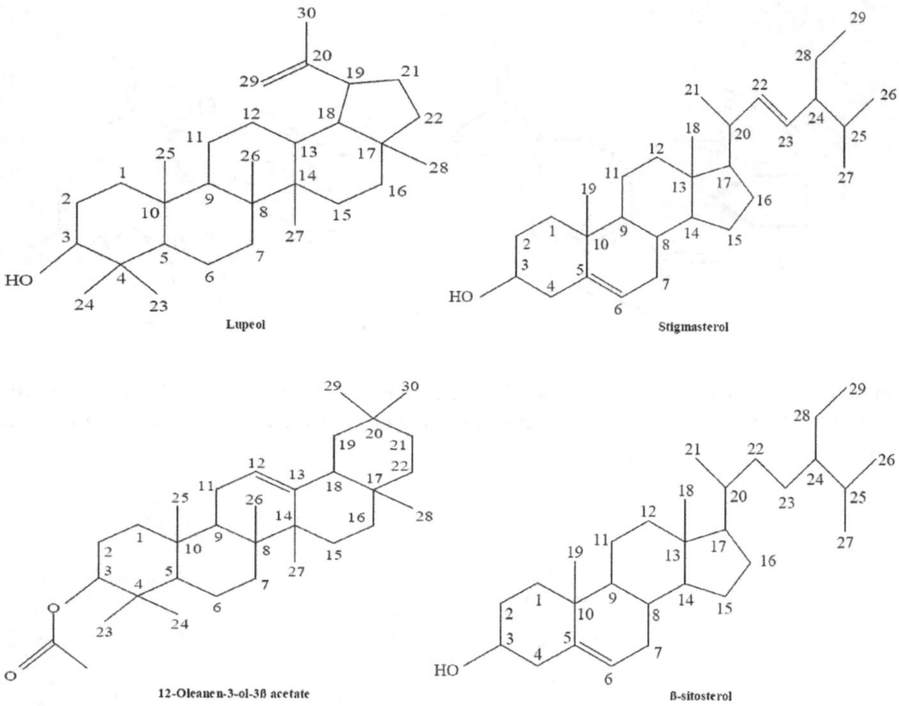

Lupeol

Stigmasterol

12-Oleanen-3-ol-3β acetate

β-sitosterol

FIGURE 3.27 Lupeol, 12-oleanen-3-ol-3β-acetate, stigmasterol, and β-sitosterol in the n-hexane portion of the plant by Haque et al. (From Haque, M.A. et al., *J. Appl. Pharm. Sci.*, 2, 79–83, 2012.)

(Figure 3.27). Bioassay-guided fractionation conducted by Youn et al. (2012) of the hexane extract from the flowers of *V. cinerea* led to the isolation of a new sesquiterpene lactone (8a-hydroxyhirsutinolide) and a new naturally occurring derivative (8a-hydroxyl-1-O-methylhirsutinolide), along with seven known compounds (Figure 3.28). The isolated compounds were evaluated for their cancer chemo-preventive potential based on their ability to inhibit nitric oxide production and tumor necrosis factor alpha (TNF-α)-induced NF-κB activity.

Soma et al. (2017) investigated the anti-malarial activity of *V. cinerea* against the *Plasmodium falciparum* strain K1, which is resistant to chloroquine, and 3D7, which is sensitive to chloroquine. The crude alkaloid extract obtained from the whole plant in this study showed good antimalarial effects against the chloroquine-resistant strain K1. In addition, moderate antimalarial effects were obtained with dichloromethane extracts against K1. Preliminary phytochemical analysis by Rizvi et al. (2011) on the methanol extract of the leaf and flower of *V. cinerea* indicated the presence of alkaloids, phenols, tannins, saponins, and flavonoids. In addition, the antibacterial activity of different extracts (hexane, petroleum ether, chloroform, and methanol) of the leaf and flower were tested separately against both Gram-positive (*S. aureus* and *Bacillus cereus*) and Gram-negative (*Enterobacter aerogenes*) bacteria using the agar well diffusion method. The antibacterial potential of the methanol

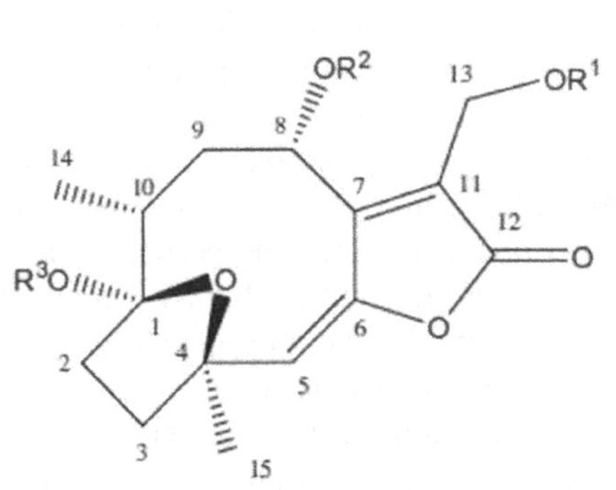

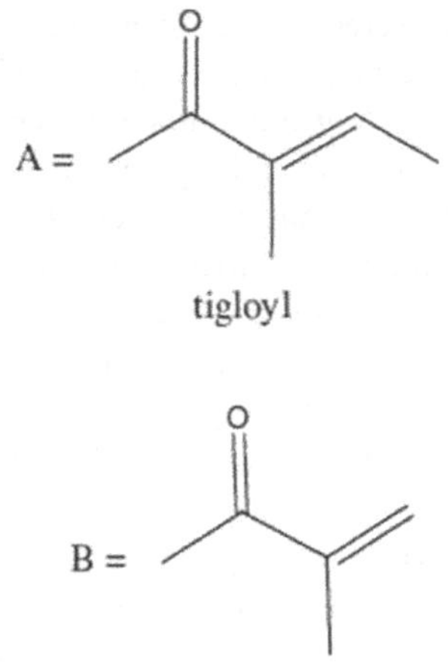

A = tigloyl

B = methylacryloyl

	R^1	R^2	R^3
1	H	A	H
2	H	H	H
3	H	H	CH_3
4	Ac	A	H
5	Ac	B	H
6	Ac	B	CH_3
7	Ac	A	CH_3
8	Ac	H	H
9	H	A	CH_3

FIGURE 3.28 Structures of compounds isolated from the hexane extract of the flower of *V. cinerea* by Youn et al. (From Youn, U.J. et al., *Bioorganic Med. Chem. Lett.*, 22, 5559–5562, 2012.)

extract of the leaf against the tested bacteria determined by minimal inhibitory concentration and minimal bactericidal concentration indicated a higher susceptibility of the *B. cereus* and *S. aureus* as compared with *E. aerogenes.* The larvicidal efficacy of *V. cinerea* leaf extracts against the common filarial vector, *Culex quinquefasciatus*, was determined by Arivoli et al. (2011). The study concluded that application of this plant's extract to larval habits may lead to promising results in filarial and mosquito management programs.

Sreedevi et al. (2011) examined the effect of petroleum ether, ethyl acetate, and alcoholic extracts of aerial parts of *V. cinerea* (500 mg/kg) on cisplatin-induced nephrotoxicity in albino rats. Out of the three extracts, the alcoholic extract showed pronounced curative activity, the ethyl acetate extract exhibited good prophylactic activity, and the petroleum ether extract showed moderate protection in both curative and prophylactic models against cisplatin-induced toxicity.

Goggi and Malpathak (2017) investigated the antioxidant activity of various crude extracts prepared from different organs of plant such as root, stem, and leaves using polar and nonpolar solvents. The antioxidant activity was determined in this study using DPPH and ABTS radical scavenging activities and a phosphor-molybdenum assay.

LC-MS screening for identification of phytoconstituents in various parts of the plant revealed the presence of isoprenoids, steroids, and phenols, which were responsible for the potent antioxidant activities. Varsha et al. (2017) evaluated the petroleum ether, ethanol, and aqueous extracts of leaves of *V. cinerea* for their antioxidant potential and anticancer activity against breast cancer cell lines. Ethanol and aqueous extracts, specifically, had antioxidant and anti-cancer properties in this study. Pratheeshkumar and Kuttan (2010) reported the protective role of *V. cinerea* against cyclophosphamide-induced toxicity in Balb/c mice. Cyclophosphamide is a widely used anti-neoplastic drug, which causes toxicity to normal cells due to its toxic metabolites. The use of cyclophosphamide in treating cancer patients is limited due to its severe toxicity induced mainly by oxidative stress. Pratheeshkumar and Kuttan (2010) reported a synergistic action of cyclophosphamide and *V. cinerea* in reducing the solid tumors in mice. Rajamurugan et al. (2011) evaluated the methanol extract of *V. cinerea* leaf for its antioxidant activity and acute toxicity in Wistar albino rats. The lethal dose (LD_{50}) of the extract was found to be greater than 2000 mg/kg, and no pathological changes were observed. In addition, the study revealed the presence of several bioactive compounds such as phenols and sesquiterpenoids in the extract, which may be responsible for its antioxidant property.

Due to its application in some traditional medicinal systems to help smoking cessation, there have been recent scientific studies systematically documenting the effects of *V. cinerea* supplementation. *V. cinerea* prepared as an herbal tea has shown smoking cessation effects in smokers (Wongwiwatthananukit et al. 2009). Leelarungrayub et al. (2010) observed in a clinical study that the herb did indeed result in a reduced smoking rate, which may be related to oxidative stress and beta-endorphin levels. Prasopthum et al. (2015) found eight active compounds isolated from *V. cinerea* that comprise inhibitory activity toward human cytochrome P450 2A6 and monoamine oxidases. These compounds were three flavones (apigenin, chrysoeriol, luteolin), one flavonol (quercetin), and four hirsutinolide-type sesquiterpene lactones (8a-(2-methylacryloyloxy)-hirsutinolide-13-O-acetate, 8a-(4-hydroxymethacryloyloxy)-hirsutinolide-13-O-acetate, 8a-tigloyloxyhirsutinolide-13-O-acetate, and 8a-(4-hydroxytigloyloxy)-hirsutinolide-13-O-acetate).

Concluding Remarks

V. cinerea remains an important herb in the Sri Lankan traditional medicinal pharmacopoeia. Although the ethnomedical applications of the plant in the country appear to be limited, it has been used for many ailments in other parts of the world where it grows. The scientific evidence is growing in support of the plant's ability to remedy diseases. It is encouraging to see that systematic biochemical and bioactive studies have been carried out on the compounds found in *V. cinerea* and are responsible for the ability of the herb to prevent certain diseases. A positive next step could be to develop these bioactive compounds – either individually or in combination – as pharmaceuticals. Clinical evidence in support of these beneficial effects needs to be produced as well. There have been no reports of lack of destructive harvesting and reproductive barriers that might hinder the growth of *V. cinerea*, which displays a robustness to withstand obstacles of habitat loss in any of the countries where it is found. This is somewhat encouraging when compared with species such as *A. lanata*.

REFERENCES

Abirami, P. and A. Rajendran. 2012. GC-MS analysis of methanol extracts of *Vernonia cinerea*. *European Journal of Experimental Biology* 2(1):9–12.

Adebajo, A.C., O.F. Ayoola, E.O. Iwalewa, A.A. Akindahunsi, N.O. Omisore, C.O. Adewunmi and T.K. Adenowo. 2006. Antitrichomonal, biochemical and toxicological activities of methanolic extract and some carbazole alkaloids isolated from the leaves of *Murraya koenigii* growing in Nigeria. *Phytomedicine* 13(4):246–254.

Adepu, A., S. Narala, A. Ganji and S. Chivalvar. 2013. A review on natural plant: *Aerva lanata*. *International Journal of Pharmaceutical Science* 3:398–402.

Adesina, S.K., O.A. Olatunji, D. Brgenthal and J. Reisch. 1988. New biogenetically significant constituents of *Clausena anisata* and *Murraya koenigii*. *Pharmazie* 43(3):221–222.

Agarwal, S.K., S.S. Singh, S. Verma, V. Lakshmi, A. Sharma and S. Kumar. 2000. Chemistry and medicinal uses of *Gymnema sylvestre* (Gurmar) leaves: A review. *Indian Drugs* 37(8):354–360.

Ahmed, A.B.A., A.S. Rao and M.V. Rao. 2010. *In vitro* callus and *in vivo* leaf extract of *Gymnema sylvestre* stimulate β-cells regeneration and anti-diabetic activity in Wistar rats. *Phytomedicine* 17:1033–1039.

Akanji, M.A., S.O. Olukolu and M.I. Kazeem. 2018. Leaf extracts of *Aerva lanata* inhibit the activities of type 2 diabetes-related enzymes and possess antioxidant properties. *Oxidative Medicine and Cellular Longevity*. doi:10.1155/2018/3439048.

Akhtar, M.S. and J. Iqbal. 1991. Evaluation of the hypoglycaemic effect of *Achyranthes aspera* in normal and alloxan-diabetic rabbits. *Journal of Ethnopharmacology* 31(1):49–57.

Akobundu, I.O. and C.W. Agyakwa. 1998. *A Handbook of West African Weeds*. Ibadan, Nigeria: International Institute of Tropical Agriculture.

Anantaworasakul, P., H. Hamamoto, K. Sekimizu and S. Okonogi. 2017a. Biological activities and antibacterial biomarker of *Sesbania grandiflora* bark extract. *Drug Discoveries and Therapeutics* 11(2):70–77.

Anantaworasakul, P., H. Hamamoto, K. Sekimizu and S. Okonogi. 2017b. *In vitro* antibacterial activity and *in vivo* therapeutic effect of *Sesbania grandiflora* in bacterial infected silkworms. *Pharmaceutical Biology* 55(1):1256–1262.

Anantaworasakul, P., S. Klayraung and S. Okonogi. 2011. Antibacterial activities of *Sesbania grandiflora* extracts. *Drug Discoveries and Therapeutics* 5(1):12–17.

Andriani, Y., N.M. Ramli, D.F. Syamsumi, M.N.I. Kassim, J. Jaafar, N.A. Suryani, L. Marlina, N.S. Musa and H. Mohamad. 2015. Phytochemical analysis, antioxidant, antibacterial and cytotoxicity properties of keys and cores part of *Pandanus tectorius* fruits. *Arabian Journal of Chemistry* (In Press – Corrected Proof). doi:10.1016/j.arabjc.2015.11.003.

Anjana, A. and J.K. Pramod. 2009. Variation in growth of *Centella asiatica* along different soil composition. *Botany Research International* 2(1):55–60.

Anusha, C.S., B.P. Kumar, H. Sini and K.G. Nevin. 2016. Antioxidant *Aerva lanata* extract suppresses proliferation and induces mitochondria mediated apoptosis in human hepatocellular carcinoma cell line. *Journal of Experimental and Integrative Medicine* 6(2):71–81.

Arivoli, S., S. Tennyson and J.J. Martin. 2011. Larvicidal efficacy of *Vernonia cinerea* (L.) (Asteraceae) leaf extracts against the filarial vector *Culex quinquefasciatus* Say (Diptera: Culicidae). *Journal of Biopesticides* 4(1):37–42.

Arora, D., M. Kumar and S.D. Dubey. 2002. *Centella asiatica*: A review of its medicinal uses and pharmacological effects. *Journal of Natural Remedies* 2(2):143–149.

Arunachalam, G., N. Subramanian, G.P. Pazhani and V. Ravichandran. 2009. Anti-inflammatory activity of methanolic extract of *Eclipta prostrata* L. (Astearaceae). *African Journal of Pharmacy and Pharmacology* 3(3):97–100.

Arunachalam, K.D., L.B. Arun, S.K. Annamalai and A.M. Arunachalam. 2015. Potential anticancer properties of bioactive compounds of *Gymnema sylvestre* and its biofunctionalized silver nanoparticles. *International Journal of Nanomedicine* 10:31–41.

Astin, J.A. 1998. Why patients use alternative medicine: Results of a national study. *Journal of the American Medicinal Association* 279(19):1548–1553.

Awad, R., D. Levac, P. Cybulska, Z. Merali, V.L. Trudeau and J.T. Arnason. 2007. Effects of traditionally used anxiolytic botanicals on enzymes of the gamma-aminobutyric acid (GABA) system. *Canadian Journal of Physiology and Pharmacology* 85(9):933–942.

Bacher, M., G. Brader, O. Hofer and H. Greger. 1999. Oximes from seeds of *Atalantia ceylanica. Phytochemistry* 50:991–994.

Bandara, B.M.R., C.M. Hewage, D.H.L.W. Jayamanne, V. Karunaratne, N.K.B. Adikaram, K.A.N.P. Bandara, M.R.M. Pinto and D.S.A. Wijesundara. 1990. Biological activity of some steam distillates from leaves of ten species of rutaceous plants. *Journal of the National Science Council of Sri Lanka* 18(1):71–77.

Behera, P.C., B.P. Bag and M. Ghosh. 2016. Anti-urolithiatic activity of hydrogenated naphthol isolated from *Aerva lanata* (L.) Juss flower extract. *Indian Journal of Traditional Knowledge* 15(3):453–459.

Behera, P.C. and M. Ghosh. 2018. Evaluation of antioxidant, antimicrobial and antiurolithiatic potential of different solvent extracts of *Aerva lanata* Linn flowers. *Pharmacognosy Magazine* 14(53):53–57.

Benson, V.L., L.M. Khachigian and H.C. Lowe. 2008. DNAzymes and cardiovascular disease. *British Journal of Pharmacology* 154:741–748.

Bhalke, R.D., M.A. Giri, S.J. Anarthe and S.C. Pal. 2010. Antiulcer activity of the ethanol extract of leaves of *Sesbania grandiflora* (Linn.). *International Journal of Pharmacy and Pharmaceutical Sciences* 2(4):206–208.

Bhandari, P.R. 2012. Curry leaf (*Murraya koenigii*) or Cure leaf: Review of its curative properties. *Journal of Medical Nutrition and Nutraceuticals* 1(2):92–97.

Bhavna, D. and K. Jyoti. 2011. *Centella asiatica*: The elixir of life. *International Journal of Research in Ayurveda and Pharmacy* 2(2):431–438.

Biswas, M., K. Biswas, A.K. Ghosh and P.K. Haldar. 2009a. A pentacyclic triterpenoid possessing analgesic activity from the fruits of *Dregea volubilis. Pharmacognosy Magazine* 5(19):90–92.

Biswas, M., P.K. Haldar and A.K. Ghosh. 2010. Antioxidant and free-radical-scavenging effects of fruits of *Dregea volubilis. Journal of Natural Science, Biology and Medicine* 1(1):29–34.

Bonde, S.D., L.S. Nemade, M.R. Patel and A.A. Patel. 2011. *Murraya koenigii* (Curry leaf): Ethnobotany, phytochemistry and pharmacology–A review. *International Journal of Pharmacy and Phytopharmocology Research* 1:23–27.

Borah, A., R.N.S. Yadav and B.G. Unni. 2011. *In vitro* antioxidant and free radical scavenging activity of *Alternanthera sessilis. International Journal of Pharmaceutical Sciences and Research* 24:1502–1506.

Bowen, I.H. and Y.N. Patel. 1987. Acridone alkaloids and other constituents of the leaves of *Atalantia ceylanica. Planta Medica* 53(1):73–75.

Bradwejn, J., Y. Zhou, D. Koszycki and J. Shlik. 2000. A double-blind, placebo-controlled study on the effects of Gotu Kola (*Centella asiatica*) on acoustic startle response in healthy subjects. *Journal of Clinical Psychopharmacology* 20(6):680–684.

Brinkhaus, B., M. Lindner, D. Schuppan and E.G. Hahn. 2000. Chemical, pharmacological and clinical profile of the East Asian medical plant *Centella asiatica. Phytomedicine* 75:427–448.

Bunpo, P., K. Kataoka, H. Arimochi, H. Nakayama, T. Kuahara, Y. Bando, K. Izumi, U. Vinitketkumnuen and Y. Ohnishi. 2004. Inhibitory effects of *Centella asiatica* on azoxymethane-induced aberrant crypt focus formation and carcinogenesis in the intestines of F344 rats. *Food and Chemical Toxicology* 42(12):1987–1997.

Bylka, W., P. Znajdek-Awiżeń, E. Studzińska-Sroka and M. Brzezińska. 2013. *Centella asiatica in cosmetology. Advances in Dermatology and Allergology* 30:46–49.

Bylka, W., P. Znajdek-Awiżeń, E. Studzińska-Sroka, A. Dańczak-Pazdrowska and M. Brzezińska. 2014. *Centella asiatica* in dermatology: An overview. *Phytotherapy Research* 28:1117–1124.

Chakraborthy, G.S. and P.M. Ghorpade. 2010. Free radical scavenging activity of *Abutilon indicum* (Linn) sweet stem extracts. *International Journal of ChemTech Research* 2(1):526–531.

Chakraborty, D.P., S.N. Ganguly, P.N. Maji, A.R. Mitra, K.C. Das and B. Weinstein. 1973. Chemical taxonomy: 32, Murrayazolinine, a carbazole alkaloid from *Murraya koenigii*. *Chemistry and Industry* 7:322–333.

Chakravarti, D. and N.B. Debnath. 1981. Isolation of gymnemagenin the sapogenin from *Gymnema sylvestre* R. Br. (Asclepiadaceae). *Journal of the Institution of Chemists* 53:155–158.

Chandrasekara, A. and F. Shahidi. 2018. Herbal beverages: Bioactive compounds and their role in disease risk reduction–A review. *Journal of Traditional and Complementary Medicine* 8(2018):451–458.

Chandrika, U.G. and P.A.A.S.P. Kumara. 2015. Gotu Kola (*Centella asiatica*): Nutritional Properties and Plausible Health Benefits. *Advances in Food and Nutrition Research* 76:125–157.

Chatterjee, T.K., A. Chakraborty, M. Pathak and G.C. Sengupta. 1992. Effects of plant extract *Centella asiatica* (Linn.) on cold restraint stress ulcer in rats. *Indian Journal of Experimental Biology* 30(10):889–891.

Chau, P.L. 2000. Ancient Chinese had their fingers on the pulse. *Nature* 404(6777):431–431.

Chauhan, P.K., I.P. Pandey and V.K. Dhatwalia. 2010. Evaluation of the antidiabetic effect of ethanolic and methanolic extracts of *Centella asiatica* leaves extract on alloxan-induced diabetic rats. *Advances in Biological Research* 4(1):27–30.

Chea, A., S. Hout, C. Long, L. Marcourt, R. Faure, N. Azas and R. Elias. 2006. Antimalarial activity of sesquiterpene lactones from *Vernonia cinerea*. *Chemical and Pharmaceutical Bulletin* 54(10):1437–1439.

China, R., S. Mukherjee, S. Sen, S. Bose, S. Datta, H. Koley, S. Ghosh and P. Dhar. 2012. Antimicrobial activity of *Sesbania grandiflora* flower polyphenol extracts on some pathogenic bacteria and growth stimulatory effect on the probiotic organism *Lactobacillus acidophilus*. *Microbiological Research* 167:500–506.

Chong, N.J. and Z. Aziz. 2015. A systematic review of the efficacy of *Centella asiatica* for improvement of the signs and symptoms of chronic venous insufficiency. *Evidence-based Complementary and Alternative Medicine*. doi:10.1155/2013/627182.

Choo, C.S.C., V. Waisundara and Y.H. Lee. 2013. Evaluation and characterization of antioxidant activity of selected herbs and spices. *Journal of Natural Remedies* 13(2):95–103.

Chopra, R.N., S.L. Nayar and I.C. Chopra. 1999. *Glossary of Indian Medicinal Plants*. New Delhi, India: National Institute of Science Communication, CSIR.

Choudhury, B.P. 1988. Assessment and conservation of medicinal plants of Bhubaneswar and its neighbourhood. In: *Indigenous Medicinal Plants*. New Delhi, India: Today & Tomorrow's Printers & Publishers, pp. 211–219.

Chowdhury, B.K. and B.C. Debi. 1969. Taxonomy XVIII, mukoeic acid; First carbazole carboxylic acid form plant sources. *Chemistry and Industry* 17:549.

Chulika, M.G.A.I., P.V.G. Chathurika, M.A.Y.R. Manchanayaka, S. Randenikumara and M.S.A. Perera. 2012. Beliefs and practices regarding herbal plants used in self-care among persons in Raththanapitiya area and the university community. *Proceedings of the Scientific Sessions*. Faculty of Medical Sciences, University of Sri Jayawardenepura, Sri Lanka.

Collins, R. 2000. *The Sociology of Philosophies: A Global Theory of Intellectual Change.* Cambridge, MA: Harvard University Press.

Cox, D.N., S. Rajasuriya, P.E. Soysa, J. Gladwin and A. Ashworth. 1993. Problems encountered in the community-based production of leaf concentrate as supplement for pre-school children in Sri-Lanka. *International Journal of Nutrition and Food Sciences* 44:123–132.

Curini, M., G. Cravotto, F. Epifano and G. Giannone. 2006. Chemistry and biological activity of natural and synthetic prenyloxycoumarins. *Current Medicinal Chemistry* 13:199–222.

Daisy, P., S. Kanakappan and M. Rajathi. 2009. Antihyperglycemic and antihyperlipidemic effects of *Clitoria ternatea* Linn. in alloxan-induced diabetic rats. *African Journal of Microbiology Research* 3(5):287–291.

Dash, G.K., P. Suresh, S.K. Sahu, D.M. Kar, S. Ganapaty and S.B. Panda. 2002. Evaluation of *Evolvulus alsinoides* Linn. for anthelmintic and antimicrobial activities. *Journal of Natural Remedies* 2:182–185.

Dashputre, N.L. and N.S. Naikwade. 2011. Evaluation of anti-ulcer activity of methanolic extract of *Abutilon indicum* Linn leaves in experimental rats. *International Journal of Pharmaceutical Sciences and Drug Research* 3(2):97–100.

Deb, D., S. Dev, A.K. Das, D. Khanam, H. Banu, M. Shahriar, A. Ashraf, M.S.K. Choudhuri and S.A.M.K. Basher. 2010. Antinociceptive, anti-inflammatory and anti-diarrheal activities of the hydroalcoholic extract of *Lasia spinosa* Linn. (Araceae) Roots. *Latin American Journal of Pharmacy* 29(8):1269–1276.

Devarai, H.K., O.S. Koushik, P.S. Babu and R. Karthikeyan. 2017. *In Vitro* anti tubercular activity of leaves of *Aerva lanata* L. *International Biology and Biomedical Journal* 3(4):209–212.

Devi, B.P., R. Boominathan and S.C. Mandal. 2003. Anti-inflammatory, analgesic and antipyretic properties of *Clitoria ternatea* root. *Fitoterapia* 74(4):345–349.

Dhanapackiam, S. and M.H.M. Ilyas. 2010. Effect of salinity on chlorophyll and carbohydrate contents of *Sesbania grandiflora* seedlings. *Indian Journal of Science and Technology* 3(1):64–66.

Di Fabio, G., V. Romanucci, A. De Marco and A. Zarrelli. 2014. Triterpenoids from *Gymnema sylvestre* and their pharmacological activities. *Molecules* 19:10956–10981.

Dinnimath, B.M., S.S. Jalalpure and U.K. Patil. 2017. Antiurolithiatic activity of natural constituents isolated from *Aerva lanata. Journal of Ayurveda and Integrative Medicine* 8(2017):226–232.

Dinnimath, M.D. and S.S. Jalalpure. 2018. Antioxidant and antiurolithiatic efficacy of *Aerva lanata* (L) fractions by *in vitro* and *in vivo* screening techniques. *Indian Journal of Pharmaceutical Education and Research* 52(3):426–436.

Djomeni, P.D.D., L. Tédong, E.A. Asongalemb, T. Dimo, D.S. Sokengc and P. Kamtchouing. 2006a. Hypoglycaemic and antidiabetic effect of root extracts of *Ceiba pentandra* in normal and diabetic rats. *African Journal of Traditional, Complementary and Alternative Medicine* 3(1):129–136.

Djomeni, P.D.D., L. Tédong, E.A. Asongalemb, T. Dimo, D.S. Sokengc and P. Kamtchouing. 2006b. Hypoglycaemic and antidiabetic effect of root extracts of *Ceiba pentandra* in normal and diabetic rats. *Indian Journal of Pharmacology* 38(3):194–197.

Dy Phon P. 2000. *Dictionnaire des plantes utilisées au Cambodge.* Phnom Penh, Cambodia: Olympic.

Ediriweera, E.R.H.S.S. and W.D. Ratnasooriya. 2009. A review on herbs used in treatment of diabetes mellitus by Sri Lankan Ayurvedic and traditional physicians. *AYU* 30:373–391.

European Medicines Agency (EMA). 2012. Science Medicines Health. http://www.ema.europa.eu (last accessed on 25 October 2018).

Fernando, C.D. and P. Soysa. 2014. Total phenolic, flavonoid contents, in-vitro antioxidant activities and hepatoprotective effect of aqueous leaf extract of *Atalantia ceylanica*. *BMC Complementary and Alternative Medicine* 14:395. doi:10.1186/1472-6882-14-395.

Fernando, C.D. and P. Soysa. 2015. Optimized enzymatic colorimetric assay for determination of hydrogen peroxide (H_2O_2) scavenging activity of plant extracts. *MethodsX* 2(2015):283–291.

Fernando, M.R., S.M.D.N. Wickramasinghe, M.I. Thabrew, P.L. Ariyananda and E.H. Karunanayake. 1991. Effect of *Artocarpus heterophyllus* and *Asteracanthus longifolia* on glucose tolerance in normal human subjects and in maturity onset diabetic patients. *Journal of Ethnopharmacology* 31:277–282.

Fraser, A.W. and J.R. Lewis. 1973. Two acridone alkaloids from *Atalantia ceylanica* (Rutaceae). *Journal of the Chemical Society – Perkin Transactions 1* 1973(17):615–616.

Gahlawat, D.J., S. Jakhar and P. Dahiya. 2014. *Murraya koenigii* (L.) Spreng: An ethnobotanical, phytochemical and pharmacological review. *Journal of Pharmacognosy and Phytochemistry* 3(3):109–119.

Gholap, S. and A. Kar. 2003. Effects of *Inula racemosa* root and *Gymnema sylvestre* leaf extracts in the regulation of corticosteroid induced diabetes mellitus: Involvement of thyroid hormones. *Pharmazie* 58:413–415.

Giri, B. and K.G. Mukerji. 2004. Mycorrhizal inoculant alleviates salt stress in *Sesbania aegyptiaca* and *Sesbania grandiflora* under field conditions: Evidence for reduced sodium and improved magnesium uptake. *Mycorrhiza* 14:307–312.

Gnanapragasam, A., K.K. Ebenezer, V. Satish, P. Govindaraju and T. Devki. 2004. Protective effect of *Centella asiatica* on antioxidant tissue defense system against adriamycin induced cardiomyopathy in rats. *Life Sciences* 16:585–597.

Gnanaraj, W.E., J.M. Antonisamy, K.M. Subramanian and S. Nallyan. 2011. Micropropagation of *Alternanthera sessilis* (L.) using shoot tip and nodal segments. *Iranian Journal of Biotechnology* 9(3):206–212.

Gobalakrishnan, R., M. Kulandaivelu, R. Bhuvaneswari, D. Kandavel and L. Kannan. 2013. Screening of wild plant species for antibacterial activity and phytochemical analysis of *Tragia involucrata* L. *Journal of Pharmaceutical Analysis* 3(6):460–465.

Goel, R.K. and K. Sairam. 2002. Anti-ulcer drugs from indigenous sources with emphasis on *Musa sapientum*, Tamrabhasma, *Asparagus racemosus* and *Zingiber officinale*. *Indian Journal of Pharmacology* 34:100–110.

Goggi, A. and N. Malpathak. 2017. Antioxidant activities of root, stem and leaves of *Vernonia cinerea* (L) Less. *Free Radicals and Antioxidants* 7(2):178–183.

Gopi, C. and T.M. Vatsala. 2006. *In vitro* studies on effects of plant growth regulators on callus and suspension culture biomass yield from *Gymnema sylvestre* R. Br. *African Journal of Biotechnology* 5(12):1215–1219.

Gothai, S., K. Muniandy, N.M. Esa, S.K. Subbiah and P. Arulselvan. 2018. Anticancer potential of *Alternanthera sessilis* extract on HT-29 human colon cancer cells. *Asian Pacific Journal of Tropical Biomedicine* 8(8):394–402.

Govindan, G., T.G. Sambandan, M. Govindan, A. Sinskey, J. Vanessendelft, I. Adenan and C.K. 2007. A bioactive polyacetylene compound isolated from *Centella asiatica*. *Planta Medica* 73(6):597–599.

Guillasper, J.N., J.M.M. Gabriel and A.M.S.D. Razalan. 2015. Pre-clinical evaluation of the aqueous extract of *Sesbania grandiflora* (katuray) as hypoglycemic agent. *BMJ Open* 5(Suppl 1):A1–A53.

Gujjeti, R.P. and E. Mamidala. 2014. Anti-HIV activity and cytotoxic effects of *Aerva lanata* root extracts. *American Journal of Phytomedicine and Clinical Therapeutics* 2(7):894–900.

Gunasekera, L. 2009. What do you know about Mukunuwenna. *The Island – Online*. http://www.island.lk/2009/10/03/satmag4.html (last accessed on 5 November 2018).

Guo, J.S., C.L. Cheng and M.W. Koo. 2004. Inhibitory effects of *Centella asiatica* water extract and asiaticoside on inducible nitric oxide synthase during gastric ulcer healing in rats. *Planta Medica* 70(12):1150–1154.

Gupta, R.S. and D. Singh. 2007. Protective nature of *Murraya koenigii* leaves against hepatosuppression through antioxidant status in experimental rats. *Pharmacologyonline* 1:232–242.

Gurav, S., V. Gulkari, N. Duragkar and A. Patil. 2007. A systemic review: Pharmacognosy, phytochemistry, pharmacology and clinical applications of *Gymnema sylvestre* R Br. *Pharmacognosy Reviews* 1:338–343.

Hamid, K., I. Ng, V.J. Tallapragada, L. Varadi, D.E. Hibbs, J. Hanrahan and P.W. Groundwater. 2016. An investigation of the differential effects of ursane triterpenoids from *Centella asiatica*, and their semisynthetic analogues, on GABA-A receptors. *Chemical Biology and Drug Design* 88(3):386–397.

Handral, H.K., A. Pandith and S.D. Shruthi. A review on *Murraya koenigii*: Multipotential medicinal plant. *Asian Journal of Pharmaceutical and Clinical Research* 5(4):5–14.

Haque, M.A., M.M. Hassan, A. Das, B. Begum, M.D. Ali and H. Morshed. 2012. Phytochemical investigation of *Vernonia cinerea* (Family: Asteraceae). *Journal of Applied Pharmaceutical Science* 2(6):79–83.

Haque, R., M.S. Ali, A. Saha and M. Allimuzzaman. 2006. Analgesic activity of methanolic extract of the leaf of *Erythrina variegata*. *Journal of Pharmaceutical Science* 5:77–79.

Hara, K., T. Someya, K. Sano, Y. Sagane, T. Watanabe and R.G.S. Wijesekara. 2018. Antioxidant activities of traditional plants in Sri Lanka by DPPH free radical-scavenging assay. *Data in Brief* 17(2018):870–875.

Hasan, N., H. Osman, S. Mohamad, W.K. Chong, K. Awang and A.S.M. Zahariluddin. 2012. The chemical components of *Sesbania grandiflora* root and their anti-tuberculosis activity. *Pharmaceuticals* 5:882–889.

Hashim, P., H. Sidek, M.H.M. Helan, A. Sabery, U.D. Palanisamy and M. Ilham. 2011. Composition and bioactivities of *Centella asiatica*. *Molecules* 16:1310–1322.

Himbutana, G.P. 2006. Ven. Thotagamuwe Sri Rahula Thera scholar monk par excellence (newspaper article). Colombo, Sri Lanka: Budu Sarana, Lake House Publishers (Retrieved 1 October 2018).

Hossain, A.I., M. Faisal, S. Rahman, R. Jahan and M. Rahmatullah. 2014. A preliminary evaluation of antihyperglycemic and analgesic activity of *Alternanthera sessilis* aerial parts. *BMC Complementary and Alternative Medicine* 14:169. doi:10.1186/1472-6882-14-169.

Hout, S., A. Chea, S.S. Bun, R. Elias, M. Gasquet, O. Timon-David, G. Balansard and N.J. Azas. 2006. Screening of selected indigenous plants of Cambodia for antiplasmodial activity. *Journal of Ethnopharmacology* 107:12–18.

Huda-Faujan, N., A. Nonhatn, A.S. Norrakiah and A.S. Babji. 2007. Antioxidant activities of water extracts of some Malaysian herbs. *ASEAN Food Journal* 14:61–68.

Iyer, U.M. and U.V. Mani. 1990. Studies on the effect of curry leaves supplementation (*Murraya koenigii*) on lipid profile, glycated proteins and amino acids in non-insulindependent diabetic patients. *Plant Foods for Human Nutrition* 40(4):275–282.

Jain, M., R. Kapadia, R.N. Jadeja, M.C. Thounaojam, R.V. Devkar and S.H. Mishra. 2011. Cytotoxicity evaluation and hepatoprotective potential of bioassay guided fractions from *Feronia limmonia* Linn leaf. *Asian Pacific Journal of Tropical Biomedicine* 1(6):443.447.

Jain, V., M. Momin and K. Laddha. 2012. *Murraya koenigii*: An updated review. *International Journal of Ayurvedic and Herbal Medicine* 2(4):607–627.

Jain, V.C., N.M. Patel, D.P. Shah, P.K. Patel and B.H. Joshi. 2010. Antioxidant and antimicrobial activities of *Alangium salvifolium* (L.F) Wang root. *Global Journal of Pharmacology* 4(1):13–18.

Jalalpure, S.S., N. Agrawal, M.B. Patil, R. Chimkode and A. Tripathi. 2008. Antimicrobial and wound healing activities of leaves of *Alternanthera sessilis* Linn. *International Journal of Green Pharmacy* 2(3):141–144.

James, J.T. and I.A. Dubrey. 2009. Pentacyclic triterpenoids from the medicinal herb, *Centella asiatica* (L.) Urban. *Molecules* 14:3922–3941.

Jana, U., T.K. Sur, L.N. Maity, P.K. Debnath and D. Bhattacharyya. 2010. A clinical study on the management of generalized anxiety disorder with *Centella asiatica*. *Nepal Medical College Journal* 12(1):8–11.

Jayanthi, G., T. Sathishkumar, T. Senthilkumar and M. Jegadeesan. 2012. Free radical scavenging potential of *Cardiospermum halicacabum* L. var. *microcarpum* (Kunth) Blume seeds. *International Research Journal of Pharmaceutical and Applied Sciences* 2(4):41–48.

Jayanthi, G., T. Sathishkumar, T. Senthilkumar and M. Jegadeesan. 2013. Effect of *Cardiospermum halicacabum* l. var. *microcarpum* (Kunth) Blume seed oil on acute and subacute inflammation. *International Journal of Pharmaceutical Research and Development* 4(12):56–59.

Jayanthi, G., S. Shahiladevi and M. Jegadeesan. 2009. Anti-inflammatory activity of *Cardiospermum halicacabum* L. var. *microcarpum* (Kunth) Blume seeds. *Hamdard Medicus* 52(2):35–37.

Jayashree, V.H. and R. Londonkar. 2014. Comparative phytochemical studies and antimicrobial potential of fruit extracts of *Feronia limonia* Linn. *International Journal of Pharmacy and Pharmaceutical Sciences* 6(1):731–734.

Jayasinghe, H.D., S.S. Rajapaksha and C. De Alwis. 2014. A compilation and analysis of food plants utilization of Sri Lankan butterfly larvae (papilionoidea). *Taprobanica* 6(2):110–131.

Jayawardena, N., M.I. Watawana and V.Y. Waisundara. 2015. Evaluation of the total antioxidant capacity, polyphenol contents and starch hydrolase inhibitory activity of ten edible plants in an *In vitro* model of digestion. *Plant Foods for Human Nutrition* 70:71–76.

Jayaweera, D.M.A. 1982. *Medicinal Plants (Indigenous and Exotic) Used in Ceylon: Part V.* Colombo, Sri Lanka: National Science Council of Sri Lanka.

Karalliadde, L. and I.B. Gawarammana. 2008. *Herbal Medicines: A Guide to its Safer Use.* London, UK: Hammersmith Press.

Karumari, R.J., S. Sumathi, K. Vijayalakshmi and S.E. Balasubramanian. 2014. Anthelmintic efficacy of *Sesbania grandiflora* leaves and *Solanum torvum* fruits against the nematode parasite *Ascaridia galli*. *American Journal of Ethnomedicine* 1(5):326–333.

Kayalvizhi, D., V. Sivakumar and M. Jayanthi. 2015. Phytochemical screening and antinephrolithiasis activity of ethanol extract of *Aerva lanata* on ethylene glycol induced renal stone in rats. *Research Journal of Pharmacy and Technology* 8(11):1481–1486.

Khan, B.A., A. Abraham and S. Leelamma. 1997. Antioxidant effect of curry leaves, *Murraya koenigii* and mustard seed, *Brassica juncea* in rats fed with high fat diet. *Indian Journal of Experimental Biology* 35:148–150.

Khare, C.P. 2007. *Indian Medicinal Plants: An Illustrated Dictionary.* Berlin, Germany: Springer Verlag.

Khramov, V.A., A.A. Spasov and M.P. Samokhina. 2008. Chemical composition of dry extracts of *Gymnema sylvestre* leaves. *Pharmaceutical Chemistry Journal* 42(1):30–32.

Kim, H.J., F. Chen, X. Wang, H.Y. Chung and Z.Y. Jin. 2005. Evaluation of antioxidant activity of vetiver (*Vetiveria zizanioides* L.) oil and identification of its antioxidant constituents. *Journal of Agricultural and Food Chemistry* 53(20):7691–7695.

Kirana, H., S.S. Agrawal and B.P. Srinivasan. 2009. Aqueous extract of *Ficus religiosa* Linn reduces oxidative stress in experimentally-induced type 2 diabetic rats. *Indian Journal of Experimental Biology* 47:822–826.

Ko, K.M., D.H. Mak, P.Y. Chiu and M.K. Poon. 2004. Pharmacological basis of 'Yang-invigoration' in Chinese medicine. *Trends in Pharmacological Sciences* 25(1):3–6.

Kohil, J.K., J.A. Patil and A.K. Gajjar. 2010. Pharmacological review on *Centella asiatica*: A potential herbal cure-all. *Indian Journal of Pharmaceutical Science* 72(5):546–556.

Kok, S.Y.Y., S.Y. Mooi, K. Ahmad and M.A. Sukari. 2012. Antitumour promoting activity and antioxidant properties of girinimbine isolated from the stem bark of *Murraya koenigii*. *Molecules* 17(4):4651–4660.

Komalavalli, N. and M.V. Rao. 2000. In vitro micropropagation of *Gymnema sylvestre*: A multipurpose medicinal plant. *Plant Cell, Tissue and Organ Culture* 61:97–10.

Kothari, S., L. Thangavelu and A. Roy. 2017. Anti-diabetic activity of *Sesbania grandiflora*: Alpha amylase inhibitory effect. *Journal of Advanced Pharmacy Education and Research* 7(4):499–502.

Krisanapun, C., P. Peungvicha, R. Temsiririrkkul and Y. Wongkrajang. 2009. Aqueous extract of *Abutilon indicum* Sweet inhibits glucose absorption and stimulates insulin secretion in rodents. *Nutrition Research* 29(8):579–587.

Krishnamoorthi, R. and K. Elumalai. 2018. *In-vitro* anticancer activity of ethyl acetate extract of *Aerva lanata* against MCF-7 cell line. *International Journal of Pharma Research and Health Sciences* 6(1):2286–2289.

Krisnamoorthy, R. 2015. Phytochemical analysis and antioxidant property of *Aerva lanata*. *International Journal of Pharmacognosy* 2(8):426–429.

Kubmarawa, D., G.A. Ajoku, N.M. Enwerem and D.A. Okorie. 2007. Preliminary phytochemical and antimicrobial screening of 50 medicinal plants from Nigeria. *African Journal of Biotechnology* 6(14):1690–1696.

Kumar, A., S. Lingadurai, A. Jain and N.R. Barman. 2010. *Erythrina variegata* Linn: A review on morphology, phytochemistry, and pharmacological aspects. *Pharmacognosy Reviews* 4(8):147–152.

Kumar, N.S., P.K. Mukharjee, S. Bhadra, B.P. Saha and B.C. Pal. 2010. Acetylcholinesterase inhibitory potential of a carbazole alkaloid, mahanimbine, from *Murraya koenigii*. *Phytotherapy Research* 24:629–631.

Kumar, R., S. Janadri, S. Kumar, D.R. Dhananjaya and S. Swamy. 2015. Evaluation of antidiabetic activity of alcoholic extract of *Sesbania grandiflora* flower in alloxan induced diabetic rats. *Asian Journal of Pharmacy and Pharmacology* 1(1):21–26.

Kumar, S., R. Kumar, A. Dwivedi and A.K. Pandey. 2014. *In Vitro* antioxidant, antibacterial, and cytotoxic activity and *in vivo* effect of *Syngonium podophyllum* and *Eichhornia crassipes* leaf extracts on isoniazid induced oxidative stress and hepatic markers. *BioMed Research International*. doi:10.1155/2014/459452.

Kumar, S.R., D. Loveleena and S. Godwin. 2013. Medicinal Property of *Murraya koenigii*–A review. *International Research Journal of Biological Sciences* 2(9):80–83.

Kumar, V.S., A. Sharma, R. Tiwari and S. Kumar. 1999. *Murraya koenigii*: A review. *Journal of Medicinal and Aromatic Plant Science* 2(1):1139–1144.

Kumaran, A. and R.J. Karunakaran. 2007. In vitro antioxidant activities of methanol extracts of five *Phyllanthus* species from India. *LWT – Food Science and Technology* 40(2):344–352.

Kundu, S., S. Chakraborty and N.N. Bala. 2018. Preliminary phytochemical investigation and in vitro free radical scavenging activity of leaves of *Sesbania grandiflora* (leguminosae). *World Journal of Pharmacy and Pharmaceutical Sciences* 7(2):1482–1488.

Kuo, Y.H., Y.J. Kuo, A.S. Yu, M.D. Wu, C.W. Ong, L.M.Y. Kuo, J.T. Huang, C.F. Chen and S.Y. Li. Two novel sesquiterpene lactones, cytotoxic vernolide-A and -B, from *Vernonia cinerea*. *Chemical and Pharmaceutical Bulletin* 51(4):425–426.

Kurihara, Y. 1969. Antisweet activity of gymnemic acid A1 and its derivatives. *Life Sciences* 8(9):537–543.

Kuroda, M., Y. Mimaki, H. Harada, H. Sakagami and Y. Sashida. 2001. Five new triterpene glycosides from *Centella asiatica*. *Natural Medicines* 55(3)134–138.

Laladhas, K.P., V.T. Cheriyan, T. Puliappadamba, S.V. Bava, R.G. Unnithan, P.L. Vijayammal and R.J. Anto. 2010. A novel protein fraction from Sesbania grandiflora shows potential anticancer and chemopreventive efficacy, *in vitro* and *in vivo*. *Journal of Cellular and Molecular Medicine* 14(3):636–646.

Latha, P.S. and K. Kannabiran. 2006. Antimicrobial activity and phytochemicals of *Solanum trilobatum* Linn. *African Journal of Biotechnology* 5(23):2402–2404.

Lee, Y.H., C. Choo, M.I. Watawana, N. Jayawardena and V.Y. Waisundara. 2014. An appraisal of eighteen commonly consumed edible plants as functional food based on their anti-oxidant and starch hydrolase inhibitory activities. *Journal of the Science of Food and Agriculture* 95:2956–2964.

Leelarungrayub, D., S. Pratanaphon, P. Pothongsunun, T. Sriboonreung, A. Yankai and R.J. Bloomer. 2010. *Vernonia cinerea* Less. supplementation and strenuous exercise reduce smoking rate: Relation to oxidative stress status and beta-endorphin release in active smokers. *Journal of the International Society of Sports Nutrition* 7:21. doi:10.1186/1550-2783-7-21.

Limmatvapirat, C., S. Sirisopanaporn and P. Kittakoop. 2004. Antitubercular and antiplasmodial constituents of *Abrus precatorius*. *Planta Medica* 70:276–278.

Lin, J.J., H. Jiang and X.T. Ding. 2017. Synergistic combinations of five single drugs from *Centella asiatica* for neuronal differentiation. *NeuroReport* 28:23–27.

Lin, S.C., C.J. Yao, C.C. Lin and Y.H. Lin. 1996. Hepatoprotective activity of Taiwan folk medicine: *Eclipta prostrata* Linn. against various hepatotoxins induced acute hepatotoxicity. *Phytotherapy Research* 10:483–490.

Liu, X., W. Ye, B. Yu, S. Zhao, H. Wu and C. Che. 2004. Two new flavonol glycosides from *Gymnema sylvestre* and *Euphorbia ebracteolata*. *Carbohydrate Research* 339(4):891–895.

Liu, X.J., B.B. Huang, J. Lin, J. Fei, Z.G. Chen, Y.Z. Pang, X.F. Sun and K.X. Tang. 2006. A novel pathogenesis-related protein (SsPR10) from *Solanum surattense* with ribonucleolytic and antimicrobial activity is stress- and pathogen-inducible. *Journal of Plant Physiology* 5(3):546–556.

Loganayaki, N., P. Siddhuraju and S. Manian. 2013. Antioxidant activity and free radical scavenging capacity of phenolic extracts from *Helicteres isora* L. and *Ceiba pentandra* L. *Journal of Food Science and Technology* 50(4):687–695.

Luqman, S., R. Kumar, S. Kaushik, S. Srivastava, M.P. Darokar and S.P.S. Khanuja. 2009. Antioxidant potential of the root of *Vetiveria zizanioides* (L.) Nash. *Indian Journal of Biochemistry and Biophysics* 46:122–125.

Malar, S., R. Manikandan, P.J.C. Favas, S.V. Sahi and P. Venkatachalam. 2014. Effect of lead on phytotoxicity, growth, biochemical alterations and its role on genomic template stability in Sesbania grandiflora: A potential plant for phytoremediation. *Ecotoxicology and Environmental Safety* 108:249–257.

Malik, A. and S. Nayak. 2011. Phytochemical and preliminary toxicity study of *Sesbania grandiflora* (Linn.) flowers. *International Journal of Biomedical and Advance Research* 2(11):444–449.

Malik, J.K., F.V. Manvi, K.R. Alagawadi and M. Noolvi. 2008. Evaluation of anti-inflammatory activity of *Gymnema sylvestre* leaves extract in rats. *International Journal of Green Pharmacy* 2(2):114–115.

Malkanthi, S.H.P., A. Karunaratne, S.D. Amuwela and P. Silva. 2014. Opportunities and challenges in cultivating underutilized field crops in Moneragala district of Sri Lanka. *Asian Journal of Agriculture and Rural Development* 4(1):96–105.

Mandal, N.A., M. Kar, S.K. Banerjee, A. Das, S.N. Upadhyay, R.K. Singh, A. Banerji and J. Banerji. 2010. Antidiarrhoeal activity of carbazole alkaloids from *Murraya koenigii* Spreng Rutaceae seeds. *Fitoterapia* 81:72–74.

Mandal, S.C. and C.K.A. Kumar. 2002. Studies on anti-diarrhoeal activity of *Ficus hispida* leaf extract in rats. *Fitoterapia* 73(7–8):663–667.

Mandal, S.C., B. Saraswathi, C.K.A. Kumar, S.M. Lakshmi and B.C. Maiti. 2000. Protective effect of leaf extract of *Ficus hispida* Linn. against paracetamol-induced hepatotoxicity in rats. *Phytotherapy Research* 14(6):457–459.

Manfred, F., M.P. John, D.S. Dajaja and A.K. Douglas. 1985. Koeniline, a further cytotoxic carbazole alkaloid from *Murraya koenigii. Phytochemistry* 24(12):3041–3043.

Matsuda, H., T. Morikawa, H. Ueda and M. Yoshikawa. 2001. Medicinal foodstuffs. XXVII. Saponin constituents of gotu kola (2): Structures of new ursane- and oleanane-type triterpene oligoglycosides, centellasaponins B, C, and D, from *Centella asiatica* cultivated in Sri Lanka. *Chemical and Pharmaceutical Bulletin* 49(10):1368–1371.

Monago, C.C. and E.O. Alumanah. 2005. Antidiabetic effect of chloroform-methanol extract of *Abrus Precatorius* Linn seed in alloxan diabetic rabbit. *Journal of Applied Science and Environmental Management* 9(1):85–88.

Mukti, M., A. Ahmed, S. Chowdhury, Z. Khatun, P. Bhuiyan, K. Debnath and M. Rahmatullah. 2012. Medicinal plant formulations of Kavirajes in several areas of Faridpur and Rajbari districts, Bangladesh. *American-Eurasian Journal of Sustainable Agriculture* 6:234–247.

Munasinghe, D.A.L., E.D.C. Karunarathna and A.D.H. Sudesh. 2015. Antibacterial activity of extract of leaves of *Atalantia ceylanica* (Yakinaran). In: *Proceedings of the International Postgraduate Research Conference 2015*, University of Kelaniya, Kelaniya, Sri Lanka, p. 163.

Muniandy, K., S. Gothai, K.M.H. Badran, S.S. Kumar, N.M. Esa and P. Arulsevlan. 2018. Suppression of proinflammatory cytokines and mediators in LPS-induced RAW 264.7 macrophages by stem extract of *Alternanthera sessilis* via the inhibition of the NF-κB pathway. *Journal of Immunology Research*. doi:10.1155/2018/3430684.

Muniandy, K., S. Gothai, W.S. Tan, S.S. Kumar, N.M. Esa, G. Chandramohan, K.S. Al-Numair and P. Arulselvan. 2018. *In Vitro* wound healing potential of stem extract of *Alternanthera sessilis. Evidence-Based Complementary and Alternative Medicine.* doi:10.1155/2018/3142073.

Murakami, N.,T. Murakami, M. Kadoya, H. Matsuda, J. Yamahara and M. Yoshikawa. 1996. New hypoglycemic constituents from 'Gymnemic Acid' from *Gymnema sylvestre. Chemical and Pharmaceutical Bulletin* 44(2):469–471.

Murashige, T. and F. Skoog. 1962. A revised medium for rapid growth and bio-assays with tobacco tissue cultures. *Physiologia Plantarum* 15:473–497.

Murugan, M. and V.R. Mohan. 2014. Phytochemical, FT-IR and antibacterial activity of whole plant extract of *Aerva lanata* (L.) Juss. Ex. Schult. *Journal of Medicinal Plant Studies* 2(3):51–57.

Muruhan, S., S. Selvaraj and P.K. Viswanathan. 2013. *In vitro* antioxidant activities of *Solanum surattense* leaf extract. *Asia Pacific Journal of Tropical Biomedicine* 3(1):28–34.

Nandagopal, S., M. Lalitha, S. Abirami, D. Sakrishna and A. Priyan. 2015. Effect of phytohormones on micropropagation and phytochemical studies of *Aerva lanata* (Linn.) Juss. ex Schult-A seasonal and vulnerable plant. *Der Pharmacia Lettre* 7(3):291–298.

Nandi, M.K., D. Garabadu, T.D. Singh and V.P. Singh. 2016. Physicochemical and phytochemical standardization of fruit of *Sesbania grandiflora. Der Pharmacia Lettre* 8(5):297–304.

Napagoda, M.T., B.M.A.S. Malkanthi, S.A.K. Abayawardana, M.M. Qader and L. Jayasinghe. 2016. Photoprotective potential in some medicinal plants used to treat skin diseases in Sri Lanka. *BMC Complementary and Alternative Medicine* 16:479. doi:10.1186/s12906-016-1455-8.

Naveen, K., C. Suresh, S.S. Kumar and T.R. Chandra. 2017. Comprehensive literature review of *mandukparni* (*Centella asiatica*) w.s.r. to its medicinal properties. *International Journal of Ayurveda and Pharma Research* 5(5):65–71.

Nayak, P., S. Nayak, D.M. Kar and P. Das. 2010. Pharmacological evaluation of ethanolic extracts of the plant *Alternanthera sessilis* against temperature regulation. *Journal of Pharmacy Research* 6(3):1381–1383.

Noreen, Y., H. El-Seedi, P. Perera and L. Bohlin. 1991. Two new isoflavones from *Ceiba pentandra* and their effect on cyclooxygenase-catalyzed prostaglandin biosynthesis. *Journal of Natural Products* 61(1):8–12.

Noviany, N., A. Nurhidayat, S. Hadi, T. Suhartati, M. Aziz, N. Purwatasari and I. Subsaman. 2018. Sesbagrandiflorain A and B: isolation of two new 2-arylbenzofurans from the stem bark of *Sesbania grandiflora*. *Natural Product Research* 32(21):2558–2564.

Nutan, M.T.H., A. Hasnat and M.A. Rashid. 1998. Antibacterial and cytotoxic activities of *Murraya koenigii*. *Fitoterapia* 69(2):173–175.

Ogawa, Y., S. Sekita, T. Umemura, M. Saito, A. Ono, Y. Kawasaki, O. Uchida, Y. Matsushima, T. Inoue and J. Kanno. 2004. *Gymnema sylvestre* leaf extract: A 52-week dietary toxicity study in Wistar rats. *Journal of the Food Hygienic Society of Japan* 45(1):8–18.

Oh, H., S.H. Lee, T. Kim, K.Y. Chai, H.T. Chung, T.O. Kwon, J.Y. Jun, O.S. Jeong, Y.C. Kim and Y.G. Yun. 2002. Furocoumarins from *Angelica dahurica* with hepatoprotective activity on tacrine-induced cytotoxicity in Hep G2 Cells. *Planta Medica* 68:463–464.

Orhan, I.E. 2012. *Centella asiatica* (L.) Urban: From traditional medicine to modern medicine with neuroprotective potential. *Evidence-based Complementary and Alternative Medicine*. doi:10.1155/2012/946259.

Othman, A., A. Ismail, F.A. Hassan, B.N.M. Yusof and A. Khatib. 2016. Comparative evaluation of nutritional compositions, antioxidant capacities, and phenolic compounds of red and green sessile joyweed (*Alternanthera sessilis*). *Journal of Functional Foods* 21:263–271.

Ou, B., D.J. Huang, M. Hampsch-Woodill and J.A. Flanagan. 2003. When east meets west: The relationship between yin-yang and antioxidation-oxidation. *The FASEB Journal* 17(2):127–129.

Pajaniradje, S., M. Kumaravel, R. Pamidimukkala, S. Subramanian and R. Rajagopalan. 2014. Antiproliferative and apoptotic effects of *Sesbania grandiflora* leaves in human cancer cells. *BioMed Research International*. doi:10.1155/2014/474953.

Pal, R.S. and Y. Pal. 2016. Pharmacognostic review and phytochemical screening of *Centella asiatica* Linn. *Journal of Medicinal Plants Studies* 4(4):132–135.

Pande, M.S., S.P.B.N. Gupta and A. Pathak. 2009. Hepatoprotective activity of *Murraya koenigii* Linn. bark. *Journal of Herbal Medicine and Toxicology* 3(1):69–71.

Panigrahi, P., C. Panda and A. Patra. 2016. Extract of *Sesbania grandiflora* ameliorates hyperglycemia in high fat diet-streptozotocin induced experimental diabetes mellitus. *Scientifica*. doi:10.1155/2016/4083568.

Pari, L. and M. Latha. 2002. Effect of *Cassia auriculata* flowers on blood sugar levels, serum and tissue lipids in streptozotocin diabetic rats. *Singapore Medical Journal* 43:617–621.

Perera, P.R.D., S. Ekanayake and K.K.D.S. Ranaweera. 2013. *In vitro* study on antiglycation activity, antioxidant activity and phenolic content of *Osbeckia octandra* L. leaf decoction. *Journal of Pharmacognosy and Phytochemistry* 2(4):158–161.

Perry, L.M. 1980. *Medicinal Plants of East and Southeast Asia*. Cambridge, MA: MIT Press.

Persaud, S.J., H. Al-Majed, A. Raman and P.M. Jones. 1999. *Gymnema sylvestre* stimulates insulin release *in vitro* by increased membrane permeability. *Journal of Endocrinology* 163:207–212.

Pithayanukul, P., S. Laovachirasuwan, R. Bavovada, N. Pakmanee and R. Suttisrib. Anti-venom potential of butanolic extract of *Eclipta prostrata* against Malayan pit viper venom. *Journal of Ethnopharmacology* 90(2–3):347–352.

Piyadasa, H.D.W. 1994. Traditional systems for preventing and treating animal diseases in Sri Lanka. *Revue Scientifique et Technique (International Office of Epizootics)* 13(2):471–486.

Pointel, J.P., M.D. Boccalon, M. Cloarec, M.D. Ledevehat and M. Joubert. 1987. Titrated extract of *Centella asiaitica* (TECA) in the treatment of venous insufficiency of the lower limbs. *Angiology* 38:46–50.

Porchezhian, E. and S.H. Ansari. 2005. Hepatoprotective activity of *Abutilon indicum* on experimental liver damage in rats. *Phytomedicine* 12(1–2):62–64.

Porchezhian, E., S.H. Ansari and S. Ahmad. 2001. Analgesic and anti-inflammatory effects of *Alangium salvifolium*. *Pharmaceutical Biology* 39(1):65–66.

Porchezhian, E. and R.M. Dobriyal. 2003. An overview on the advances of *Gymnema sylvestre*: Chemistry, pharmacology and patents. *Pharmazie* 58(1):5–12.

Potawale, S.E., V.M. Shinde, L. Anandi, S. Borade, H. Dhalawat and R.S. Deshmukh. 2008. *Gymnema sylvestre*: A comprehensive review. *Pharmacologyonline* 2:144–157.

Prakash, V., N. Jaiswal and M. Srivastava. 2017. A review on medicinal properties of *Centella asiatica*. *Asian Journal of Pharmaceutical and Clinical Research* 10(10):69–74.

Prasopthum, A., P. Pouyfung, S. Sarapusit, E. Srisook and P. Rongnoparut. 2015. Inhibition effects of *Vernonia cinerea* active compounds against cytochrome P450 2A6 and human monoamine oxidases, possible targets for reduction of tobacco dependence. *Drug Metabolism and Pharmacokinetics* 30(2):174–181.

Pratheeshkumar, P. and G. Kuttan. 2010. Ameliorative action of *Vernonia cinerea* L. on cyclophosphamide-induced immunosuppression and oxidative stress in mice. *Inflammopharmacology* 18:197–207.

Pratumvinit, B., T. Srisapoomi, P. Worawattananon, N. Opartkiattikul, W. Jiratchariyakul and T. Kummalue. 2009. *In vitro* antineoplastic effect of *Ficus hispida* L. plant against breast cancer cell lines. *Journal of Medicinal Plants Research* 3(4):255–261.

Priya, C.L., G. Kumar, L. Karthik and K.V.B. Rao. 2010. Antioxidant activity of *Achyranthes aspera* Linn stem extracts. *Pharmacologyonline* 2:228–237.

Punturee, K., C.P. Wild and U. Vinitketkumneun. 2004. Thai medicinal plants modulate nitric oxide and tumor necrosis factor-alpha in J774.2 mouse macrophages. *Journal of Ethnopharmacology* 95:183–189.

Rahman, M.A., S.C. Bachar and M. Rahmatullah. 2010. Analgesic and antiinflammatory activity of methanolic extract of *Acalypha indica* Linn. *Pakistan Journal of Pharmaceutical Sciences* 23(3):256–258.

Rahmatullah, M., A.A. Mahmud, M.A. Rahman, M.F. Uddin, M. Hasan, M.A. Khatun, A.B.M.A. Bashar et al. 2011. An ethnomedicinal survey conducted amongst folk medicinal practitioners in the two southern districts of Noakhali and Feni, Bangladesh. *American-Eurasian Journal of Sustainable Agriculture* 5:115–131.

Raja, M., T. Umapoorni and N. Kanya. Antidiabetic activity of *Aerva lanata* Linn Juss by using alloxan induced diabetic rats. *International Journal of Pharma Research and Health Sciences* 5(4):1775–1778.

Rajamurugan, R., N. Selvaganabathy, S. Kumaravel, CH. Ramamurthy, V. Sujatha, M. Suresh Kumar and C. Thirunavukkarasu. 2011. Identification, quantification of bioactive constituents, evaluation of antioxidant and *in vivo* acute toxicity property from the methanol extract of *Vernonia cinerea* leaf extract. *Pharmaceutical Biology* 49(12):1311–1320.

Rajanna, L., C. Nagaveni and M. Ramakrishnan. 2011. *In vitro* shoot multiplication of a seasonal and vulnerable medicinal plant–*Aerva lanata* L. *International Journal of Botany* 7(3):255–259.

Rajasekaran, R. and Y. Gebrekidan. 2018. A review on antibacterial phytochemical constitutions present in *Aerva lanata* and their mode of action against bacterial biofilm. *International Journal of Pharmaceutical & Biological Archives* 9(1):16–30.

Rajendran, M.P., B.B. Pallaiyan and N. Selvaraj. 2014. Chemical composition, antibacterial and antioxidant profile of essential oil from *Murraya koenigii* (L.) leaves. *Avicenna Journal of Phytomedicine* 4(3):200–214.

Rajiv, P. and R. Sivaraj. 2012. Screening for phytochemicals and antimicrobial activity of aqueous extract of *Ficus religiosa* Linn. *International Journal of Pharmacy and Pharmaceutical Sciences* 4(5):207–209.

Ramesh, T., R. Mahesh and V.H. Begum. 2007. Effect of *Sesbania grandiflora* on lung antioxidant defense system in cigarette smoke-exposed rats. *International Journal of Biological Chemistry* 1(3):141–148.

Ramesh, T., C. Sureka, S. Bhuvana and V.H. Begum. 2010. *Sesbania grandiflora* diminishes oxidative stress and ameliorates antioxidant capacity in liver and kidney of rats exposed to cigarette smoke. *Journal of Physiology and Pharmacology* 61(4):467–476.

Ramnath, V., G. Kuttan and R. Kuttan. 2002. Immunpotentiating activity of abrin, a lectin from *Abrus precatorius* Linn. *Indian Journal of Experimental Biology* 40:910–913.

Ramsewak, R.S., M.G. Nair, G.M. Strasburg, D.L. Dewitt and J.L. Nitiss. 1999. Biologically active carbazole alkaloids from *Murraya koenigii*. *Journal of Agricultural and Food Chemistry* 47:444–447.

Ranaweera, S.S. 1996. Mosquito larvicidal activity of some Sri Lankan plants. *Journal of the National Science Council of Sri Lanka* 24(2):63–70.

Rao, L.J.M., K. Ramalakshmi, B.B. Borse and B. Raghavan. 2006. Chemical composition of volatiles from coconut sap neera and effect of processing. *Food Chemistry* 100:742–747.

Rao, P. 2018. Ayurvedic view of *Alternanthera Sessilis* Linn. with special reference to Mathsyakshi: A brief review. *SciFed Journal of Herbal Medicine* 2:1.

Rastogi, R.P. and B.N. Mehrotra. 1980. *Compendium of Indian Medicinal Plants*, Volume 2. New Delhi, India: Central Drug Research Institute and National Institute of Science Communication, pp. 473–475.

Reddy, B.T., D. Ali Moulali, E. Anjaneyulu, M. Ramgopal, K. Hemanth Kumar, O. Lokanatha, M. Guruprasad and M. Balaji. 2010. Antimicrobial screening of the plant extracts of *Cardiospermum halicacabum L.* against selected microbes. *Ethnobotanical Leaflets* 14:911–919.

Reddy, J.S., P.R. Rao and M.S. Rao. 2002. Wound healing effects of *Heliotropium indicum*, *Plumbago zeylanicum* and *Acalypha indica* in rats. *Journal of Ethnopharmacology* 79(2):249–251.

Reddy, P.S., G.R. Gopal and G.L. Sita. 2004. *In vitro* multiplication of *Gymnema sylvestre* R.Br., an important medicinal plant. *Current Science* 10(4):1–4.

Reisch, J., A.C. Adebajo, A.J. Aladesanmi, K.S. Adesina, D. Bergenthal and U. Meve. 1994. Chemotypes of *Murraya koenigii* growing in Sri Lanka. *Planta Medica* 60:295–296.

Reisch, J., A.C. Adebazo, V. Kumar and A.J. Aladesanmi. 1994. Two carbazole alkaloids from *Murraya koenigii*. *Phytochemistry* 36(4):1073–1076.

Reuther, W. and H.J. Webber. 1967. *The Citrus Industry*. Berkeley, CA: University of California, Division of Agricultural Sciences.

Riya, M.P., K.A. Antu, S. Pal, A.K. Srivastava, S. Sharma and K.G. Raghu. Nutraceutical potential of *Aerva lanata* (L.) Juss. ex Schult ameliorates secondary complications in streptozotocin-induced diabetic rats. *Food and Function* 5:2086–2095.

Rizvi, S.M.D., D. Biswas, J.M. Arif and M. Zeeshan. 2011. *In-vitro* antibacterial and antioxidant potential of leaf and flower extracts of *Vernonia cinerea* and their phytochemical constituents. *International Journal of Pharmaceutical Sciences Review and Research* 9(2):164–169.

Rumalla, C.S., Z. Ali, A.D. Weerasooriya, T.J. Smillie and I.A. Khan. 2010. Two new triterpene glycosides from *Centella asiatica*. *Planta Medica* 76(10):1018–1021.

Sabaragamuwa, R., C.O. Perera and B. Fredrizzi. 2018. *Centella asiatica* (Gotu kola) as a neuroprotectant and its potential role in healthy ageing. *Trends in Food Science and Technology* 79:88–97.

Saima, Y., A.K. Das, K.K. Sarkar, A.K. Sen Sr. and P. Sur. An antitumor pectic polysaccharide from *Feronia limonia*. *International Journal of Biological Molecules* 27(5):333–335.

Samuelsson, G. 1992. *Drugs of Natural Origin. A Textbook of Pharmacognosy*. Stockholm, Sweden: Swedish Pharmaceutical Press.

Samy, R.P., P. Gopalakrishnakone, M. Sarumathi and S. Ignacimuthu. 2006. Wound healing potential of *Tragia involucrata* extract in rats. *Fitoterapia* 77(4):300–302.

Sancho, R., N. Marquez, M. Gomez-Gonzalo, M.A. Calzado, G. Bettoni, M.T. Coiras, J. Alcami, M. Lopez-Cabrera, G. Appendino and E. Munoz. 2004. Imperatorin inhibits HIV-1 replication through an Sp1-dependent pathway. *Journal of Biological Chemistry* 279:37349.

Sane, R.T., V.V. Kuber, M.S. Chalissery and S. Menon. 1995. Hepatoprotection by *Phyllanthus amarus* and *Phyllanthus debilis* in CCl_4-induced liver dysfunction. *Current Science* 68(12):1243–1246.

Saneja, A., C. Sharma, K.R. Aneja and R. Pahwa. 2010. *Gymnema Sylvestre* (Gurmar): Areview. *Der Pharmacia Lettre* 2(1):275–284.

Sathasivam, P. and T. Lakshmi. 2017. Brine shrimp lethality assay of *Sesbania grandiflora* ethanolic extract–*in vitro* study. *Journal of Advanced Pharmacy Education and Research* 7(1):28–30.

Savage, K., J. Firth, C. Stough and J. Sarris. 2018. GABA-modulating phytomedicines for anxiety: A systematic review of preclinical and clinical evidence. *Phytotherapy Research* 32:3–18.

Selvanayagam, Z.E., S.G. Gnanavendhan, P. Chandrasekhran, K. Balakrishna and R.B. Rao. 1995. Plants with antisnake venom activity: A review on pharmacological and clinical studies. *Fitoterapia* 65:99–111.

Semwal, B.C., M. Verma, Y. Murti and H.N. Yadav. 2018. Neuroprotective activity of *Sesbania grandifolara* seeds extract against celecoxib induced amnesia in mice. *Pharmacognosy Journal* 10(4):747–752.

Sen, S.K. and L.M. Behera. 2008. Ethnomedicinal plants used by the tribals of Bargarh district to cure diarrhea and dysentery. *Indian Journal of Traditional Knowledge* 7:425–428.

Senadheera, S.P.A.S. and S. Ekanayake. 2012. Green leafy porridges: How good are they in controlling glycaemic response? *International Journal of Food Science and Nutrition* 64:169–174.

Senadheera, S.P.A.S. and S. Ekanayake. 2013. Development of an herbal leafy porridge from *Scoparia dulcis*. *Agricultural Sciences* 4(9B):81–84.

Senadheera, S.P.A.S., S. Ekanayake and C. Wanigatunge. 2014. Anti-diabetic properties of rice-based herbal porridges in diabetic Wistar rats. *Phytotherapy Research* 28(10):1567–1572.

Senadheera, S.P.A.S., S. Ekanayake and C. Wanigatunge. 2015. Anti-hyperglycaemic effects of herbal porridge made of *Scoparia dulcis* leaf extract in diabetics–a randomized crossover clinical trial. *BMC Complementary and Alternative Medicine* 15:410. doi:10.1186/s12906–015-0935-6.

Sertié, J.A.A., G. Wiezel, R.G. Woisky and J.C.T. Carvalho. 2001. Antiulcer activity of the ethanol extract of *Sesbania grandiflora*. *Brazilian Journal of Pharmaceutical Sciences* 37(1):109–112.

Shahjahan, M., K.E. Sabitha, M. Jainu and C.S.S. Devi. 2004. Effect of *Solanum trilobatum* against carbon tetrachloride induced hepatic damage in albino rats. *Indian Journal of Medical Research* 120:194–198.

Shanmugarajan, T.S., M. Arunsundar, I. Somasundaram, E. Krishnakumar, D. Sivaraman and V. Ravichandran. 2008. Cardioprotective effect of *Ficus hispada* Linn on cyclophosphamide provoked oxidative myocardial injury in a rat model. *International Journal of Pharmacology* 4(2):78–87.

Shigematsu, N., R. Asano, M. Shimosaka and M. Okazaki. 2001. Effect of administration with the extract of *Gymnema sylvestre* R. Br leaves on lipid metabolism in rats. *Biological and Pharmaceutical Bulletin* 24(6):713–717.

Shimizu, K., A. Iino, J. Nakajima, K. Tanaka, S. Nakajyo, N. Urakawa, M. Atsuchi, T. Wada and C. Yamashita. 1997. Suppression of glucose absorption by some fractions extracted from *Gymnema sylvestre* leaves. *Journal of Veterinary Medical Science* 59(4):245–251.

Shirwaikar, A., R.H.N. Ashwatha and P. Mohapatra. 2006. Antioxidant and antiulcer activity of aqueous extract of polyherbal formulation. *Indian Journal of Experimental Biology* 44:474–480.

Siddharaju, P., A. Abirami, G. Nagarani and M. Sangeethapriya. 2014. Antioxidant capacity and total phenolic content of aqueous acetone and ethanol extract of edible parts of *Moringa oleifera* and *Sesbania grandiflora. International Journal of Biological, Biomolecular, Agricultural, Food and Biotechnological Engineering* 8(9):1090–1098.

Siddiqui, B.S., H. Aslam, S.T. Ali, S. Khan and S. Begum. 2007. Chemical constituents of *Centella asiatica. Journal of Asian Natural Products Research* 9(4):407–414.

Siddiqui, Y., T.M. Islam, Y. Naidu and S. Meon. 2011. The conjunctive use of compost tea and inorganic fertiliser on the growth, yield and terpenoid content of *Centella asiatica* (L.) urban. *Scientia Horticulturae* 130:289–295.

Singh, A., T. Kandasamy and B. Odhav. 2009. *In vitro* propagation of *Alternanthera sessilis* (sessile joyweed), a famine food plant. *African Journal of Biotechnology* 8(21):5691–5695.

Singh, S., A. Gautam, A. Sharma and A. Batra. 2010. *Centella asiatica* (l.): A plant with immense medicinal potential but threatened. *International Journal of Pharmaceutical Sciences Review and Research* 4(2):9–17.

Sinha, R. 2018. Nutritional analysis of few selected wild edible leafy vegetables of tribal of Jharkhand, India. *International Journal of Current Microbiology and Applied Sciences* 7(2):1323–1329.

Siripurapu, K.B., P. Gupta, G. Bhatia, R. Maurya, C. Nath and G. Palit. 2005. Adaptogenic and anti-amnesic properties of *Evolvulus alsinoides* in rodents. *Pharmacology Biochemistry and Behaviour* 81(3):424–432.

Solangaarachchi, S.M. and B.M.S. Perera. 1993. Floristic composition and medicinally important plants in the understory of the tropical dry mixed evergreen forest at the Hurulu Reserve of Sri Lanka. *Journal of the National Science Council of Sri Lanka* 21(2):209–226.

Soma, A., S. Sanon, A. Gansané, L.P. Ouattara, N. Ouédraog, J.B. Nikiema and S.B. Sirima. 2017. Antiplasmodial activity of *Vernonia cinerea* Less (Asteraceae), a plant used in traditional medicine in Burkina Faso to treat malaria. *African Journal of Pharmacy and Pharmacology* 11(5):87–93

Somchit, M.N., R.A. Rashid, A. Abdullah, A. Zuraini, Z.A. Zakaria, M.R. Sulaiman, A.K. Arifah and A.R. Mutalib. 2010. *In vitro* antimicrobial activity of leaves of *Acalypha indica* Linn. (Euphorbiaceae). *African Journal of Microbiology Research* 4(20):2133–2136.

Sondhi, N., R. Bhardwaj, S. Kaur, M. Chandel, N. Kumar and B. Singh. 2010. Inhibition of H_2O_2-induced DNA damage in single cell gel electrophoresis assay (comet assay) by castasterone isolated from leaves of *Centella asiatica. Health* 2(6):595–602.

Song, T.E. and L.L. Chin. 1991. Carotenoid composition and content of Malaysian vegetables and fruits by the AOAC and HPLC Methods. *Food Chemistry* 41:309–339.

Sreedevi, A., K. Bharathi and K.V.S.R.G. Prasad. 2011. Effect of *Vernonia cinerea* aerial parts against cisplatin-induced nephrotoxicity in rats. *Pharmacology Online* 2:548–555.

Sreelakshmi, R., P.G. Latha, M.M. Arafat, S. Shyamal, V.J. Shine, G.I. Anuja, S.R. Suja and S. Rajasekaran. 2007. Anti-inflammatory, analgesic and anti-lipid peroxidation studies on stem bark of *Ficus religiosa* Linn. *Natural Product Radiance* 6(5):377–381.

Srivastava, R. and Y.N. Shukla. 1996. Some chemical constituents from *Centella asiatica. Indian Drugs* 33(5):233–234.

Stice, E., S. Yokum and J.M. Gau. 2017. Gymnemic acids lozenge reduces short-term consumption of high-sugar food: A placebo controlled experiment. *Journal of Psychopharmacology* 31(11):1496–1502.

Subban, R., A. Veerakumar, R. Manimaran, K.M. Hashim and I. Balachandran. 2008. Two new flavonoids from *Centella asiatica* (Linn.). *Journal of Natural Medicines* 62(3):369–373.

Suganya, R. and G. Subasri. 2018. Phytochemical characterization and evaluation of antimicrobial activity of *Aerva lanata* L. *World Journal of Science and Research* 3(1):23–29.

Suntornsuk, L. and O. Anurukvorakun 2005. Precision improvement for the analysis of flavonoids in selected Thai plants by capillary zone electrophoresis. *Electrophoresis* 26(3):648–660.

Tachibana, Y., H. Kikuzaki, N.H. Lajis and N. Nakatani. 2001. Antioxidative activity of carbazoles from *Murraya koenigii* leaves. *Journal of Agricultural and Food Chemistry* 49:5589–5594.

Tachibana, Y., H. Kikuzaki, N.H. Lajis and N. Nakatani. 2003. Comparison of anti oxidative properties of carbazole alkaloids from *Murraya koenigii* leaves. *Journal of Agricultural and Food Chemistry* 51:6461–6467.

Tada, Y., Y. Shikishima, Y. Takaishi, H. Shibata, T. Higuti, G. Honda, M. Ito, Y. Takeda, O.K. Kodzhimatov, O. Ashurmetov and Y. Ohmoto. 2002. Coumarins and gamma-pyrone derivatives from Prangos pabularia: Antibacterial activity and inhibition of cytokine release. *Phytochemistry* 59:649–654.

Tembhurne, S.V. and D.M. Sakarkar. 2010. Beneficial effects of ethanolic extract of *Murraya koenigii* Linn. leaves in cognitive deficit aged mice involving possible anticholinesterase and cholesterol lowering mechanism. *International Journal of PharmTech Research* 2(1):181–188.

Tewtrakul, S., S. Subhadhirasakul, S. Cheenpracha and C. Karalai. 2007. HIV-1 protease and HIV-1 integrase inhibitory substances from *Eclipta prostrata. Phytotherapy Research* 21(11):1092–1095.

Thabrew, M.I., P.D.T.M. Joice and W. Jayatissa. 1987. A comparative study of the efficacy of *Pavetta indica* and *Osbeckia octandra* in the treatment of liver dysfunction. *Planta Medica* 53(3):239–241.

Thangaiyan, K. and S. Ramanujam. 2017. Isolation, identification and anti-microbial studies of Baicalein-7-*O*-glucuronide from *Gymnema sylvestre. Papirex – Indian Journal of Research* 6(4):582–584.

Thangavel, A., S. Balakrishnan, A. Arumugam, S. Duraisamy and S. Muthusamy. 2014. Phytochemical screening, gas chromatography-mass spectrometry (GC-MS) analysis of phytochemical constituents and anti-bacterial activity of *Aerva lanata* (L.) leaves. *African Journal of Pharmacy and Pharmacology* 8(5):126–135.

Tiwari, P., B.N. Mishra and N.S. Sangwan. 2014. Phytochemical and pharmacological properties of *Gymnema sylvestre*: An important medicinal plant. *BioMed Research International*. doi:10.1155/2014/830285.

Tomar, A. 2017. *Aerva lanata* (Linn.) Juss. use to cure headache. *Journal of Medicinal Plants Studies* 5(2):329–330.

Ueno, M. 1993. The bioactivity and use of the sugar absorption inhibitor *Gymnema sylvestre. Technical Journal on Food Chemistry and Chemicals* 12:21–26.

Ullah, M.O., S. Sultana and A. Haque. 2009. Antimicrobial, cytotoxic and antioxidant activity of *Centella asiatica. European Journal of Scientific Research* 30(2):260–264.

Upadhyay, S.K., A. Saha, B.D. Bhatia and K.S. Kulkami. 2002. Evaluation of the efficacy of Mentat in children with learning disability: A placebo-controlled double-blind clinical trial. *Neurosciences Today* 3:184–188.

Vani, M., S.A. Rahaman and A.P. Rani. 2017. *In vivo* antiasthmatic studies and phytochemical characterization on the stem extracts of *Alternanthera sessilis* L. using Guinea pigs model. *Journal of Entomology and Zoology Studies* 5(2):1160–1171.

Vani, M., S.A. Rahaman and A.P. Rani. 2018. Detection and quantification of major phytochemical markers for standardization of *Talinum portulacifolium, Gomphrena serrata, Alternanthera sessilis* and *Euphorbia heterophylla* by HPLC. *Pharmacognosy Journal* 10(3):439–446.

Varsha, V., S.N. Suresh and V. Prejeena. 2017. Phytochemical screening, antioxidant property and anticancer potential against mcf-7 cell lines of *Vernonia cinerea* leaves. *International Journal of Recent Advances in Multidisciplinary Research* 4(2):2335–2341.

Vetrichelvan, T. and M. Jegadeesan. 2003. Effect of alcohol extract of *Achyranthes aspera* Linn. on acute and subacute inflammation. *Phytotherapy Research* 17(1):77–79.

Wagh, V.D., K.V. Wagh, Y.N. Tandale and S.A. Salve. 2009. Phytochemical, pharmacological and phytopharmaceutics aspects of *Sesbania grandiflora* (Hadga): A review. *Journal of Pharmacy Research* 2(5):889–892.

Waisundara, V. and Y.H. Lee. 2014. A comparative study on the antioxidant activity of commonly used south Asian herbs. *Journal of Traditional and Complementary Medicine* 3(4):263–267.

Waisundara, V.Y. 2018. Assessment of bioaccessibility: A vital aspect for determining the efficacy of superfoods. In: *Current Topics in Superfoods*, N. Shiomi (Ed.). Rijeka, Croatia: InTech Open.

Waisundara, V.Y. and M.I. Watawana. 2014. The classification of Sri Lankan medicinal herbs: An extensive comparison of the antioxidant activities. *Journal of Traditional and Complementary Medicine* 4(3):196–202.

Wang, C.C., J.E. Lai, L.G. Chen, K.Y. Yen and L.L. Yang. 2000. Inducible nitric oxide synthase inhibitors of Chinese herbs. Part 2: Naturally occurring furanocoumarins. *Bioorganic Medicinal Chemistry Letters* 8:2701–2707.

Wang, X.S., J.Y. Duan and J.N. Fang. 2004. Structural features of a polysaccharide from *Centella asiatica*. *Chinese Chemical Letters* 15(2):187–190.

Wei, T.S., B.J. Geng, L.H. Qi, G.K. Tiong, K.S. Chi and W.W. Hoong. 2018. Effect of bleaching using sodium hydroxide on pulp derived from *Sesbania grandiflora*. *Journal of Tropical Resource and Sustainable Science* 6(2018):1–3.

Weragoda, P. 1980. The traditional system of medicine in Sri Lanka. *Journal of Ethnopharmacology* 2(1):71–73.

Wongwiwatthananukit, S., P. Benjanakaskul, T. Songsak, S. Suwanamajo and V. Verachai. 2009. Efficacy of *Vernonia cinerea* for smoking cessation. *Journal of Health Research* 23:31–36.

Yapa, P.A. 2015. '*Kandata Upan Thurulatha*' (Sinhala Text). Colombo, Sri Lanka: Samayawardhana Publishers Private Limited.

Yoshida, M., M. Fuchigami, T. Nagao, H. Okabe, K. Matsunaga, J. Takata, Y. Karube et al. 2005. Anti-proliferative constituents from umbelliferae plants VII. Active triterpenes and rosmarinic acid from *Centella asiatica*. *Biological and Pharmaceutical Bulletin* 28(1):173–175.

Yoshikawa, K., K. Amimoto, S. Arihara and K. Matsuura. 1989a. Structure studies of new antisweet constituents from *Gymnema sylvestre*. *Tetrahedron Letters* 30(9):1103–1106.

Yoshikawa, K., K. Amimoto, S. Arihara and K. Matsuura. 1989b. Gymnemic acid V, VI and VII from gur-ma, the leaves of *Gymnema sylvestre* R. Br. *Chemical and Pharmaceutical Bulletin* 37(3):852–854.

Yoshitaka, T. and N. Van Ke. 2007. *Edible Wild Plants of Vietnam: The Bountiful Garden*. Bangkok, Thailand: Orchid Press.

Youn, U.J., E.J. Park, T.P. Kondratyuk, C.J. Simmons, R.P. Borris, P. Tanamatayarat, S. Wongwiwatthananukit et al. 2012. Anti-inflammatory sesquiterpene lactones from the flower of *Vernonia cinerea*. *Bioorganic and Medicinal Chemistry Letters* 22:5559–5562.

Zahan, R., M.B. Alam, M.S. Islam, G.C. Sarker, N.S. Chowdhury, S.B. Hosain, M.A. Mosaddik, M. Jesmin and M.E. Haque. 2011. Anti-cancer activity of *Alangium salvifolium* flower in Ehrlich ascites carcinoma bearing mice. *International Journal of Cancer Research* 7(3):254–262.

Zhang, X.P., C.M. Wu, H.F. Wu et al. 2013. Anti-hyperlipidemic effects and potential mechanisms of action of the caffeoylquinic acid-rich *Pandanus tectorius* fruit extract in hamsters fed a high fat-diet. *PLoS ONE* 8(4):e61922. doi:10.1371/journal.pone.0061922.

Zhang, Y., X.L. Li, W.P. Lai, B. Chen, H.K. Chow, C.F. Wu, N.L. Wang, X.S. Yao and M.S. Wong. 2007. Anti-osteoporotic effect of *Erythrina variegata* L. in ovariectomized rats. *Journal of Ethnopharmacology* 109(1):165–169.

Zhao, Y., L. Peng, W. Lu, Y.Q. Wang, X.F. Huang, C. Gong, L. He, J.H. Hong, S.S. Wu and X. Jin. 2015. Effect of *Eclipta prostrata* on lipid metabolism in hyperlipidemic animals. *Experimental Gerontology* 62:37–44.

List of Medicinal Herbs

Abrus precatorius
Abutilon indicum
Acalypha indica
Achyranthes aspera
Aerva lanata
Alangium salvifolium
Allium sativum
Aloe vera
Alpinia galangal
Alternanthera denticulate
Alternanthera philoxeroides
Alternanthera sessilis
Asparagus racemosus
Asteracantha longifolia
Atlantia ceylanica
Azadirachta indica
Capsicum frutescens
Cardiospermum microcarpum
Carum copticum
Cassia auriculata
Ceiba pentandra
Centella asiatica
Cissampelos pareira
Clitoria ternatea
Coccinia grandis
Corypha umbraculifera
Costus speciosus
Croton macrostachyus
Dregea volubilis
Eclipta alba
Eclipta prostrata
Erythrina variegate
Evolvulus alsinoides
Feronia limonia
Ficus hispida
Ficus religiosa
Hemidesmus indicus
Inula racemosa
Kalanchoe laciniata

Lasia spinosa
Mimosa pudica
Murraya koenigi
Ocimum tenuiflorum
Osbeckia octandra
Pandanus tectorius
Phyllanthus debilis
Plectranthus amboinicus
Salvia officinalis
Scoparia dulcis
Sesbania grandiflora
Solanum surattense
Solanum trilobatum
Sphaeranthus indicus
Syngonium podophyllum
Tragia involucrata
Vernonia cinerea
Vetiveria zizanioides
Vitex negundo

Index

Note: Page numbers in italic and bold refer to figures and tables, respectively.

Milton Keynes UK
Ingram Content Group UK Ltd.
UKHW040053071024
449327UK00019B/538